General Contractor Business Model for Smart Cities

I dedicate this work to my family who have supported me in both the good times and the bad. Above all, I bestow it to my wife who cheerfully sacrificed time, energy and bread so that I could fulfill my dream. Her infallible love and absolute backing allowed me to get to the shore when I was drowning in doubt. Not forgetting my four, beautiful little angels – my raison d'être

I would also like to mention my parents, as well as my brothers and sisters, and their respective families; I am truly grateful for all the encouragement you have given me and the love you have shown me. Though it has been a long road, together, we have made it through. No words could truly express how appreciative I am. This success is for you – you were part of it – you always will be – now and in the future

General Contractor Business Model for Smart Cities

Elie Karam

iSTE

WILEY

First published 2022 in Great Britain and the United States by ISTE Ltd and John Wiley & Sons, Inc.

ISTE Ltd
27-37 St George's Road
London SW19 4EU
UK

www.iste.co.uk

John Wiley & Sons, Inc.
111 River Street
Hoboken, NJ 07030
USA

www.wiley.com

Library of Congress Control Number: 2021950884

British Library Cataloguing-in-Publication Data
A CIP record for this book is available from the British Library
ISBN 978-1-78630-790-3

FSC www.fsc.org MIX Paper from responsible sources FSC® C013604

Contents

List of Acronyms

AIIB	Asian Infrastructure Investment Bank
AMT	Additive Manufacturing Technology
ART	Augmented Reality Technology
BIM	Building Information Modeling
BM	Business Model
BMI	Business Model Innovation
BT	Business Technology
CCTV	Closed-Circuit TeleVision
CLD	Causal Loop Diagram
CM	Construction Manager
CPLC	Construction Project LifeCycle
CVP	Customer Value Proposition
DSR	Design Science Research
EU	European Union

GC	General Contractor
GDP	Gross Domestic Product
GHG	GreenHouse Gas
ICT	Information and Communication Technology
IDI	In-Depth Interview
IoT	Internet of Things
IPCC	Intergovernmental Panel on Climate Change
IS	Information System
IT	Information Technology
LCA	LifeCycle Assessment
MIS	Management Information System
MSP	MultiSided Platform
PM	Project Manager
R&D	Research and Development
ROI	Return On Investment
TLC	Technology LifeCycle
U-	Ubiquitous-
UK	United Kingdom
UN	United Nations
UNESCO	United Nations Educational, Scientific and Cultural Organization
USA	United States of America

USD	United States Dollar
VRT	Virtual Reality Technology
WW	World War
WWW	World Wide Web

Preface

The WWW (World Wide Web) – the Internet – and the ICT (Information and Communication Technology) – are all key, fairly recent and successive inventions that have substantially altered and shaped the world we live in (as we know it!). They have turned our existence upside down and affected every aspect of our lives, even the tiniest ones. Having said that, let us now reconsider how life is organized on Earth. To do this, let us simply think of the planet as an inhabited human body (an interlocked system). It is the transit system of roads and railways, bridges and tunnels, as well as air and seaports that enable our mobility across the continents – much like the vascular system that powers the human body. It is also the oil and gas pipelines and electricity grids that distribute energy and ensure the unblemished work of the nervous system of communications; additionally, it is the Internet cables, satellites, mobile networks and data centers, which allow for the smooth exchange and storage of information. Like the human body, the main components of the Earth system are interconnected by flows of energy and materials. And so we believe that a disruption of any of those flows would unquestionably sway the system in its entirety.

Today, this ever-growing infrastructural system – *connectography* as labeled by Parag Khanna[1] – consists of several million kilometers of roads, railways, pipelines and Internet cables. It represents a quantum leap in the mobility of people, goods, resources, knowledge and ideas. It is an evolution of the world from *political geography* (how we legitimately divide the world) to *functional geography* (how

1 This Preface was inspired by Mr. Khanna's Ted Talk: Khanna, M. (2016). How megacities are changing the map of the world? [Online]. Available at: https://www.ted.com/talks/parag_khanna_how_megacities_are_changing_the_map_of_the_world/discussion?; anguage=en [Accessed March 31, 2021].

we actually use the world). It is hence quite evident that *connectivity* – rather than *sovereignty* – has developed into the organizing principle of the human species.

"Our yearly spending on global infrastructure is anticipated to rise to USD 9 trillion within the coming decade," Mr. Khanna indicated. "We will build more infrastructure in the next forty years than we have in the past four thousand years."

Thus far, Asia is topping the list of continents whose countries are investing the most time and money into promoting their connectivity, both regionally and internationally. In collaboration with some adjacent countries, China, for instance, announced in 2015 the creation of the AIIB (Asian Infrastructure Investment Bank), a multilateral development bank that aims to support the building of infrastructure in the Asia-Pacific region. Beyond connectivity, the 21st century seems to be characterized by a second megatrend: *planetary urbanization*. According to Parag Khanna, "over 67% of the world's population will live in cities by 2030 – megacities in specific" – a justly rational argument, as megacities are nowadays on the rise and could easily be dotted anywhere in the world:

– the technology powerhouse of Silicon Valley (USA) is a very good example of a megacity extending from San Francisco, going south through Palo Alto, all the way down to St Jose;

– the sprawl of Los Angeles is another good example of a megacity, spreading south all the way to San Diego, crossing the Mexican border, to finally reach Tijuana. San Diego to Tijuana is a binational conurbation, comprising over 5 million residents as well as a joint airport terminal;

– one additional example of a megacity is the USA's northeastern megalopolis – stretching from Boston to New York, to Philadelphia and Washington – the so-called *Bos-Wash corridor*. The latter is the second most populous megacity in the USA, with over 50 million residents.

However, the megacity trend looks like a viral phenomenon; Asia remains by far the area with the biggest megacities in the world:

– from Tokyo, to Nagoya, to Osaka stretches the world's largest megacity. It comprises over 80 million people and accounts for most of Japan's economy;

– China's megacities seem to be on the rise as well, as clusters are coming together with populations reaching 100 billion people. The Yangtze River Delta, for instance, which is a triangle-shaped megacity cluster, covers an area of about 100,000 square kilometers and is home to over 115 million people (as of 2013). In 2018, the Yangtze River Delta had a GDP (Gross Domestic Product) of about USD 2.2 trillion – roughly the same size as Italy today.

These facts are weighty – also to some extent amusing – especially when we picture global diplomatic institutions (e.g. the Group of Twenty) basing their memberships on economic size rather than national representation. Under a similar scenario, some Chinese megacities would be granted access and have seats at the table, while whole countries like Argentina or Indonesia would see their partaking being revoked. The exact same leaning (*towards increased connectivity*) could be found in other countries, say India (Delhi), Iran (Tehran) and Egypt (Cairo-Alexandria corridor). And there is Lagos too – Africa's largest city in Nigeria's commercial hub – with its plans to create a rail network that would make it the anchor of a vast Atlantic coastal corridor – stretching across Benin, Togo and Ghana, to Abidjan (in Côte d'Ivoire). In other words, in some parts of the world, whole and entire countries could in time become suburbs of megacities – a plausible setup in a megacity world.

Going forward, it is worth noting that people normally move to cities to be connected, and connectivity is why these cities ultimately prosper. Whether it is São Paulo, Istanbul or Moscow (really!), any one of them has a GDP approaching or exceeding 33–50% of their entire national GDP. Bringing up the case of Gauteng province in South Africa – comprising Johannesburg and Pretoria (the capital) – it too accounts for more than 33% of the country's GDP. Equally importantly, the latter is also home to the offices of almost every single multinational that directly invests in South Africa and (circuitously) the entire African continent.

As-is, planetary urbanization seems to be a good thing, a promising megatrend. Yet, we ask, is it *risk-free*? For some, urbanization is a source of negative externalities that would lead to frustration in the long run. In their opinion, urbanized cities are destroying the planet – and will continue to do so in the future. Hitherto, today, there are over 200 intercity learning networks booming, focusing on a single goal: *sustainable urbanization* – and having a lone objective: *upholding the well-being of people*. Fair talk indeed, nevertheless, could we put our faith in such upmarket promises? Yes, we can. Looking into the matter from a different perspective, we may ask ourselves the following question:

– Do we really believe that developed nations, through summits held recurrently, would eventually succeed in reducing GHG (Greenhouse Gas) emissions and stop climate change? No, we do not.

We could reverse global warming by injecting sulfur into the stratosphere – an unconventional solution to an exceptional problem. Yet, until now (thank goodness!), there has been no need for such eccentric tenacities, especially since human beings have started to mitigate the carbon intensity of their respective economies via intercity handovers of technology, knowledge and policies. That is, that cities used to be part of the problem, but now, they are part of the solution. What is more, if we travel through megacities from end-to-end, one could easily notice extreme disparities within the same geography – another serious challenge for sustainable urbanization – and still, our global stock of financial assets has never been greater, approaching 300 trillion dollars. That is four times the actual GDP of the world. Indeed, since the latest financial crisis, we have taken on some huge debts, but – sadly! – did not invest them in *inclusive growth*. Therefore, it is only when sufficient and affordable public housing projects are built and robust investments in transportation networks are made that alienated cities and societies will come to feel complete again.

According to Parag Khanna, "connectivity is an opportunity — one of the most important asset classes of the present century". Besides *connectivity* and *equitability*, megacities could also make the world more peaceful. How? By looking at regions of the world with dense relations across borders, we see only trade and investment trails, as well as stability. Following World War 2, once industrial integration had kicked off, it in due course led to the rise of the EU (European Union). In North America, the most important streaks on the map are not the USA–Canada or USA–Mexico borders, but the dense network of roads and railways, pipelines and electricity grids, as well as water canals.

Now, let us go back to Asia – Southeast Asia in particular. This region of 600 million people is evolving into the so-called *Pax Asiana* – a period of peace among Southeast Asian nations. A similar phenomenon is taking place in East Africa where six or so countries are investing in inter-nation railways and corridors so that noncoastal countries can get their commodities to the marketplace. At last, we wonder whether connectivity could overcome the patterns of rivalry among the great powers. And to amply answer this question, what would be better than to look at the experience of East Asian countries in this respect? After all, this is the region where WW3 was supposed to break out.

China and Japan, on the one hand, have had a long history of rivalry, often deploying their air forces and navies to show off their strengths in island disputes. Then, just some time ago, Japan started making large investments in China – Japanese cars are selling big in China – and guess where the largest number of foreigners residing in Japan comes from? *BOOM*, you guessed it: China. China and India, on the other hand, have also fought a major war and have three outstanding border disputes, but today, India is the second largest shareholder in the AIIB. The two countries are currently working together to build a corridor distending from Northeast India, through Myanmar and Bangladesh, to Southern China – and their trade volumes have grown from USD 20 billion a decade ago to USD 80 billion today.

We end with these words by Parag Khanna: “connectivity has remarkably developed into a new reality – a reality that has allowed cities and nations to aggregate over time into more diplomatic and well-off wholes”. “Though no one could swear for sure today that World War 3 will not break out, anyone could realize why it has not happened yet.”

Elie KARAM
October 2021

Introduction

When we try to pick out anything by itself, we find it hitched to everything else in the Universe.

John MUIR

Connectivity has become a reality today – and the upsurge of megacities at the international level is not expected to cease anytime soon. Neither is the flow of the world population leaving for urban centers. In fact, UN data[1] shows that over half of the world's population lived in cities in 2015, and this figure is likely to rise by an extra 10% in 2030. As previously evoked, people tend to migrate to large urban centers in search of opportunities and connectivity. And these same reasons – we suppose – will continue to drive this kind of migration towards big cities. Consequently, those big cities would eventually capture a significant share of the world's wealth and their giant potential would attract further newcomers. With that said, it is valid to enquire whether today's big cities are all equally and sufficiently equipped, urbanized and structured enough to receive superfluous inhabitants (the answer is obviously: No, not at all!). Also, we may ask: *what would happen if an increase in population ends up being unaccompanied by a comparable increase in economic performance*? (normally, the quality of life of inhabitants would plunge!).

Tokyo (Japan), Delhi (India), Shanghai (China), São Paolo (Brazil) and Mexico City (Mexico) are the world's most densely populated megacities today – with respective populations exceeding 22 million. Lagos (Nigeria), Kinshasa (Democratic Republic of the Congo), Dar Es-Salaam (Tanzania) and Bombay (India) are megacities of the future; currently in the making. However, strange as this may

1 United Nations, Department of Economic and Social Affairs, Population Division (2018). World Urbanization Prospects: The 2018 Revision [Online]. Available at: https://esa.un.org/unpd/wup/Publications [Accessed March 31, 2021].

sound, urbanization has proved to be a contagious phenomenon that is spreading fairly swiftly across the world (Schaffers *et al.* 2011). Overall, this phenomenon (of urbanization) will eventually defy any nation, from the perspective of basic goods and services, and with minimum infrastructure required – a challenge that could not be resolved except through satisfactory innovation: *the creation of smart cities*[2].

Though the notion of smart cities is not new, it is the managerial approach that policy makers, city governors, mayors and project owners frequently opt for when building smart cities that could make the whole difference, mainly by rendering smart cities – smart again. Often, smart city projects are conceived and built following a *top-down approach* with the aim of improving the places we, the people, live in, yet they repeatedly fail to hit the target and don't reach envisioned goals (Turok 2014). Why? Conventionally – and mistakenly – construction companies used to (and sometimes still do!) overweigh the significance of technology, data and cutting-edge computing, while disregarding the foremost component of any successful smart city: *people*. To be sure, for a smart city to succeed and reach its full potential, it should not solely focus on the technology or the infrastructure. Instead, it must be about, must reflect the needs and wants of, and must be built for – *the people*. Otherwise, *the people* will reject it. When building smart cities, by choosing a *people-centered approach*, we inspire collective thinking, the exchange of ideas as well as the democratization of the development of cities, and instigate the city-as-a-platform concept. Consequently, projects would then be conceived with a thorough understanding of real city problems – following a *bottom-up approach*, exactly as opined by city stakeholders: *citizens*, *businesses* and *visitors*.

We add that human-centric smart cities can only be planned and developed when citizens have the opportunity to make their voice heard by governments regarding plausible methods that could be implemented to better city operations. This concept goes farther than thinking of the citizen as a source of data – but as a source of new ideas too (Neirotti *et al.* 2014). Definitely, engaging members of the public early in the conception–construction process could help eradicate disapproval when smart projects or initiatives are implemented. Smart cities formerly focused on connecting infrastructure for better insights, but the attention is nowadays shifting bit-by-bit towards engaging governments, citizens and businesses with the objective of providing upgraded city services and a higher quality of life (i.e. enhancing the citizen experience). That is, smart city projects are now deemed successful if and only if they are accepted and validated by the people – this is a prerequisite.

While the motivation of cities remained intact over time, that is, the founding of livable environments, where people and businesses could thrive together – the setups used to this end evolved favorably. Data is now put in the hands of both end-users

2 Ibid.

and policy makers to drive better decision-making, and collective thinking and intelligence are laying the groundwork for the creation of hands-on solutions to some of the toughest urban snags encountered today.

Within this framework, it is worth mentioning that a number of cities (Barcelona, London, etc.) have already started – only a few years ago – to upgrade their infrastructure systems via the implementation of sensor technology and data analytics. They are doing so in an effort to lift the performance of their physical infrastructures and the use-value of their urban assets: public transit, waste management services, wastewater systems and roads (among others). The *SmartSantander* project in Santander, Spain (Hernandez-Munoz and Munoz 2013) – an extensive real-world experimental facility, spreading across four European countries (Spain, Germany, the UK and Serbia) – is a great example of a human-centric smart initiative. Truthfully, as big cities around the globe will, in the future, be expected to shelter superfluous people, smart initiatives and enablers like the *SmartSantander* framework have been growing in strength and gaining momentum in recent years, hence underlining the need for better and more efficient methods for the management and development of big cities overall.

I.1. Relevance

Hurried urbanization puts an incredible amount of pressure on urban centers, presenting challenges for cities to provide economic opportunities and environmental sustainability, and ensure the safety, protection and well-being of their inhabitants (Moir *et al.* 2014). Prosperous cities tend to overcome these challenges by seeking sustainable and resilient growth. The IoT (Internet of Things) in smart cities – and smart city technology – account for only part of the solution. The other part, however, has to do with *adaptability*: the adoption of a new business model that allows for the optimization of the construction industry's value chain. This is valid as the conception and construction of smart real estate, often grasped as megaprojects, take years to be completed, *up to 20 years*. Whereas the TLC (Technology Lifecycle) is relatively shorter, *up to 15 years*, which signifies that there is a high risk that the implemented technology would reach its tipping point – and so become obsolete – even before the construction project has been finalized or delivered to clients (Kordas *et al.* 2015).

On another note, it is worth indicating that businesses as much as governments are facing the challenges of urban growth – and their capacity to drive continued growth is being put to the question. Actually, for businesses to be able to attract the educated talent desired, they need (at least at first) their head offices to be based in livable (and blooming) cities (Nam and Pardo 2011), meaning that the attractiveness of cities is a function (among other factors) of their respective economic power and influence – which means that talented people would not choose to migrate to big cities unless they have a good reason to do so. In this sense, cities are responding by finding ways (adequate platforms, business models, frameworks, structures or ecosystems) to speed up construction works and improve decision-making processes, not only by consulting with governmental entities – but with businesses and residents too, thus tapping into the collective intelligence of the city in its entirety (Termeer and Bruinsma 2016).

I.2. Importance

Big cities (or megacities) have long been engines of economic growth and opportunity. A World Bank analysis[3] of 750 cities around the globe found that, from 2005 through to 2012, economic growth in almost three-fourths of cities outpaced their respective national economies. By 2025, the world's top 600 cities are estimated to account for more than half of global GDP. London today accounts for almost 20% of the UK's GDP. In the USA, the Bos-Wash corridor – and the Los Angeles metropolitan area – account for nearly 35% of the country's GDP[4]. The world is now seeing a nonstop concentration of population in cities. And cities that are not suitably equipped to handle growth (or fail to adapt to this new reality) are likely to see their environments and residents suffering from negative consequences. Now, this challenge is becoming increasingly relevant as quite a few big cities around the world (e.g. Lagos, Bombay, etc.) are undergoing such an explosive growth.

To recap, a smart city is not only about implementing smarter things (conversely to what is believed today!); it is also about endowing stakeholders (policy makers, residents, project owners, project managers, etc.) with the right tools and frameworks so they can make smarter decisions. For example, in Amman, Jordan, the city authorities have lately adopted a data-driven approach to streamlining the waste management process, which allowed for the successful optimization of existing fleet management systems[5]. Not only in Amman, but also all over the globe,

3 The World Bank Group (2015). Competitive Cities for Jobs & Growth: What, Who, & How? [Online] Available at: http://documents.worldbank.org/curated/en/902411467990995484/pdf/101546-REVISED-Competitive-Cities-for-Jobs-and-Growth.pdf [Accessed April 5, 2021].

4 CityLab (2014). The Dozen Regional Powerhouses Driving the U.S. Economy [Online]. Available at: https://www.citylab.com/life/2014/03/dozen-regional-powerhouses-driving-us-economy/8575/ [Accessed April 5, 2021].

5 ASCIMER (2014). Facing the challenges of a new era: Smart city projects [Online]. Available at: http://eiburs-ascimer.transyt-projects.com/files/14_MaqousiAli_Presentation_%20%5BASCIMER%5D.pdf [Accessed April 5, 2021].

cities are adopting shrewd managerial approaches to building their own smart cities. All of this sounds great, yet we note that there is currently no unique, matchless and unrivaled smart-city model that countries around the world could use – at any time, in any way – to refurbish or build their own. And that the conception and development of smart cities is almost certain to vary from one country to another, based on each country's specificities and available resources. Furthermore, any smart development, we presume, involves numerous key players and it is the interplay between these players throughout the lifespan of the construction project that determines whether the latter will eventually succeed or fail (Ke *et al.* 2015; Komninos *et al.* 2015). This in fact sheds light on the key role of the *general contractor* who brings all stakeholders involved in construction projects together, hence tying up the construction value chain, downstream to upstream.

I.3. The managerial question

This book revolves around five major themes. From the most wide-ranging to the more specific they are as follows: *connectivity*, *smart cities*, *smart constructions*, *general contractors* and *business models*. Indeed, the main puzzle that we aim to solve can be written as follows:

– How can we design a *general contractor* business model in order to build smart cities?

Thus, as a means to an end, we assume that the most successful smart city projects (in France in particular and around the world in general) are run, managed and built by general contractors, in partnership with other construction actors. As a result, the value chain of the entire construction industry – we expect – can be optimized through improved connectivity between various stakeholders and, ultimately, so can the livability of big cities and the quality of life of residents in general.

(These assumptions are still to be validated in due course of this book!)

I.4. Outline of the book

This book follows a funnel structure and is split into six chapters, as follows:

– Chapter 1 focuses on smart cities. An extensive overview of smart city concepts and theories is proposed. That is, a full review of the existing literature on smart cities is presented, coupled with a full description of relevant theoretical models, notions and concepts. A historical synopsis of the emergence of the concept of smart cities is also given. Moving forward, some detailed, real-life examples of successful smart cities around the globe are examined. Global smart-development best practices are depicted too – in an attempt to inform an expert managerial business model on building spot-on smart cities.

– Chapter 2 explains the key roles that general contractors could play in building smart cities.

– Chapter 3 embarks upon business model theories. The features, constituents and importance of business models are methodically examined. Additionally, light is shed on recent literature relating to business model innovations.

– Chapter 4 describes the business model design process, primarily how our idea of a new business model developed over time, and how it cultivated year-after-year to finally turn into something tangible, a dual-use business model, at the end of this book.

– Chapter 5 classifies the main problems faced in French construction today and suggests plausible solutions for their resolve. Specifically, using the triple layered business model canvas, problems and solutions are dispersed across the different dimensions of a construction project: *economic*, *social* and *environmental*. To these, you may notice, we add an extra one: the *technical* dimension. In addition, the links between problems and solutions are highlighted, narratively and visually through dependency graphs and construction process maps. Additionally, an in-depth explanation of the research findings is proposed throughout this chapter. The main discussions revolve around topics such as the need for a central operator to lead smart developments, the potential benefits of a new business model in construction and the opinions of key stakeholders apropos the envisioned business model and the role of general contractors in improving the configuration and enactment of the construction industry's value chain. Lastly, Chapter 5 elucidates how innovation is part of construction processes on various levels – and how innovative procedures in general are likely to soar over time from augmented managerial and technical connectivity.

– Chapter 6 discusses our theoretical and methodological contributions to the literature on smart cities, general contractors and business models.

– Finally, we summarize all this research by putting forward some recommendations for both industry players and policy makers as to what could be done to better crack smart developments' inherent problems, irrespective of whether those problems are of a technical and/or legal nature. Hence, we call for all stakeholders to join forces to embolden the development of smart cities – and render them – literally – smart again. Research limitations and avenues are also marked in the conclusion.

1

On Smart Cities: A Literature Review

Thus far, we have established – with a fair amount of confidence – that *functional geography* has gradually come to prevail, overshadowing the significance and governance associated with *political geography*. We have also highlighted that *global urbanization* is putting serious pressure on big cities in terms of their ability to accommodate additional inhabitants, and that this pressure is being passed onto the construction industry which is currently hastening developments: *smart cities*. Furthermore, we have shed light on the importance of promoting macro-level connectivity as a means of boosting economic growth and sponsoring environmental sustainability and political stability[1].

Besides macro-level connectivity, we endeavor in this work to demonstrate that connectivity is critical at the micro-level too (Hanna 2009). Clearly, there is now a need to institute strong micro-level connectivity (within the construction industry) and rethink existing stakeholder management approaches, for the most part to improve stakeholder engagement throughout the various stages of the CPLC (Construction Project Lifecycle). In fact, establishing strong links between construction actors – maintaining open channels of communication at all times – has become a prerequisite to successful construction projects. This is palpable – in our opinion – because stakeholders often have distinct interests and concerns, and therefore strong connections at the micro-level – if they could exist – would ensure timely, consistent, relevant exchanges of information between them. The result is the successful execution of smart development projects.

1 The Smart City in 2030. First Workshop on Network 2030, Otthein Herzog. New York City, October 2018. Available at: https://www.itu.int/en/ITU-T/Workshops-and-Seminars/201810/Documents/Ottein_Herzog_Presentation.pdf [Accessed April 7, 2021].

However, connections at the micro-level connectivity cannot be established without the intervention of a key industry player who is both a connector and a maven: the *general contractor*.

In summary, this chapter explains the smart city concept in detail.

The research is then explored further to address the crucial role of the general contractor in managing the construction value chain (Chapter 2).

From our perspective, the general contractor role exists because of the presence of (a major issue) broad organizational gaps and weaknesses within the construction industry's value chain, which often hold construction actors back from proposing value-adding innovations to their clients. As for the solution to this particular issue, we declare it is nowhere to be found except in a newly-designed – centralized – stage-based business model.

Over the course of my professional career as a *general contractor*, I have had the opportunity to handle numerous construction projects and amass sufficient knowledge about the current state of the market: its strengths, weaknesses, the threats faced and future prospects to be seized. In my personal view, smart cities are likely to shape the future of urban developments. The smart city market in France, as well as in other countries, has lots to offer and has been growing exponentially, especially with the global urbanization trend which has erupted in recent years and has been on the rise ever since.

Practically, smart city projects refer to long-term construction projects that are generally labor-, capital- and technology-intensive. To say the least, they are a source of value, creating various revenue streams that could potentially benefit all construction actors[2].

Currently, the number of construction actors – those intervening at different stages of the construction value chain in order to execute smart city projects – is significant; however, the difficulty encountered in this context has very little to do with the number of those involved. Instead, it stems from the absence of an entrepreneurial entity that could resourcefully orchestrate these interventions, avoid overlapping roles and guarantee the successful delivery of projects (Flyvbjerg and Holm 2002; Flyvbjerg *et al.* 2003).

---

2 Xiong (2018). Cost–benefit analysis of smart cities technologies and applications. A thesis submitted to the Faculty of the University of Delaware. Available at: http://udspace.udel.edu/bitstream/handle/19716/23818/Xiong_udel_0060M_13359.pdf?sequence=1&isAllowed=y [Accessed April 7, 2021].

1.1. Historical synopsis

The concern surrounding sustainable development of urban centers has been a central preoccupation for many years, and the aspects that typify the cities of tomorrow have been embraced during this time. Furthermore, the vocabulary associated with the features of these cities has been unequivocally enhanced over the past few decades, mainly to explain the substantial number of concepts endorsed by stakeholders (Eremia *et al.* 2017). Today, this vocabulary has changed once again, with specific terms gaining or losing ground over time (see Tables 1.1 and 1.2).

Domain	Social	Economic	Governing
Garden cities	Participative cities	Entrepreneurial cities	Managed cities
Sustainable cities	Walkable cities	Competitive cities	Intelligent cities
Eco-cities	Integrated cities	Productive cities	Product cities
Green cities	Inclusive cities	Innovative cities	Efficient cities
Compact cities	Just cities	Business-friendly cities	Well-led cities
Smart cities	Open cities	Global cities	Smart cities
Resilient cities	Livable cities	Resilient cities	Future cities

Table 1.1. *The cities of tomorrow, conception of success (source: adapted from Guerrero-Pérez* et al. *(2013))*

Post-1950, *sustainable city* was the most popular English term used to label future urban developments. *Digital city* followed in the late 1990s; its popularity resulting from its inherent ability to connect with and reflect the increasing importance of ICT at the time. Nevertheless, in 2009, the term gradually fell out of favor and was replaced by a new one: *smart city*. This new term embraces elements of sustainability and social inclusion, while at the same time being suited to the evolution of the IoT (Deakin 2012). In the following sections, we shed light on the two concepts of *sustainable* and *smart cities*. Then, in the chapters that follow, we embark on deep dive into the main features and models of smart cities and present a quasi-complete inventory of all definitions of smart cities.

Term	Popularity	Regional Popularity	Popularity in Countries	Popularity in Cities
Future cities	Stable	Global	India, USA, Canada, Australia, UK, Mexico, Brazil	Minneapolis, Singapore, Mumbai, New Delhi, Phoenix, San Francisco, Pune
Eco-cities	Stable	Asia	The Philippines, Singapore, Malaysia, India	Chandigarh, Tianjin
Smart cities	Fluctuating interest	Europe, North America	Italy, Spain, Belgium, UK	Barcelona, Bologna, Turin, Rome
Intelligent cities	Stable	North America	USA, UK	London
Sustainable cities	Stable	Commonwealth	Australia, UK, Canada, USA, India	Vancouver, Singapore, Washington, Auckland, Portland, Dubai, London, Austin
Compact cities	Stable	Mixed	Australia, UK, USA	Salt Lake City, New York City
Liveable cities	Rarely used	Commonwealth	Australia, UK, Canada, Singapore	New York City, Singapore, Melbourne, Pittsburgh, Vancouver
Digital cities	Stable, following a drop in popularity	Mixed	USA, Ireland, The Philippines, UK	Kansas City, Oklahoma City, Dublin, Minneapolis
Innovative cities	Stable	Mixed	USA, UK, India	Bangalore
Green cities	Stable	Northern America	USA, Australia, Canada	New York City

Table 1.2. *Geographic trends in future city term usage (source: adapted from Moir* et al. *(2014))*

1.1.1. *Sustainability, unfolded*

The idea of sustainability grew out of the cognizance that the prevailing model (at the time) of socioeconomic development was oblivious to the ecological disasters created by industrialized economies, as well as to mounting social disparities among people and nations (Najam 2005). As of 1970, growing concern over humanity's influence on the environment ignited a global environmental movement that culminated in the United Nations Conference on the Human Environment, also known as the Stockholm Conference. Initially, this conference was met with much resistance from the emerging countries represented, who scrutinized the lawfulness of environmental issues as a global priority. The statement from Côte d'Ivoire demonstrates this conviction pretty well: "more pollution problems are better than more poverty problems, as far as they are proof of industrialization" (Rowland 1973). As described by Najam (2005), global environmental debates are still today very much a legacy of industrialized versus emerging economies.

After Stockholm, the debates have shifted from solely focusing on environmental problems to (collectively) concentrating on environmental and socioeconomic development, now recognized as the concept of *sustainable development* (Prizzia 2007). The Brundtland Report[3] describes sustainable development as "development that meets the needs of the present without compromising the ability of future generations to meet their own needs" (WCED 1987). Since it first appeared on the world stage, the concept of sustainable development has garnered widespread attention and has become the key message when promoting the environment in the fields of science, technology, economy and urban planning (Bibri and Krogstie 2017).

As is it easy to confuse the concepts of *sustainable development* and *sustainability* with each other, it seems vital at this point to elucidate that the latter describes a long-term goal or sought-after end-state that could be preserved over time, while the former describes initiatives and projects undertaken to attain the ultimate goal of sustainability (Höjer and Wangel 2015).

UNESCO[4] defines sustainability as "a paradigm for thinking about the future in which environmental, societal and economic concerns are balanced in the pursuit of an improved quality of life". According to Elkington (1997), sustainability is

3 *Our Common Future*, that is, *The Brundtland Report*, was published in 1987. Targeting multilateralism and interdependence of nations in the search for a "sustainable development" path, it sought to recapture the spirit of the Stockholm Conference which had introduced environmental concerns to the formal political development sphere.

4 Sustainable Development. Available at: https://en.unesco.org/themes/education-sustainable-development/what-is-esd/sd [Accessed April 10, 2021].

commonly described as an equilibrium of its three dimensions – economic, social and environmental – and is also referred to as the triple-bottom-line perspective of people, planet and prosperity. Currently, the most prominent school of thinking within the European sustainability logic is ecological modernization (Baker *et al.* 1997), the basic aim of which is to replace a manufacturing-based economy with one that is cleaner and more service-oriented (Pepper 1998). This includes (but is not limited to) improvement in energy and resource efficiency.

In the 21st century, the work carried out by scientists at the IPCC (Intergovernmental Panel on Climate Change) has helped to uphold the significance and consideration given to the environmental aspects of sustainability on the global stage. Similarly, their strident warnings against business-as-usual impacts on the environment have created a global call for a transition to sustainability. Today, sustainability is certainly the end goal in the face of unconventional challenges (Rittel and Webber 1973), and a plausible solution to particularly pressing issues, for example, climate change.

1.1.2. *The sustainable in a city*

As the world's population is becoming increasingly urban (refer to the Preface and Introduction), sustainable development is now perhaps the most fundamental archetype influencing research and dialogue around urban development, affording it a high level of recognition within global policy (De Jong *et al.* 2015). The sustainable city concept became largely dominant in Europe via the Aalborg Charter (1994)[5], an urban sustainability initiative, comprising over 3,000 signatories today, making it the single most successful European effort in sustainable urban development. Its goal is evident: "to render urban settlements inclusive, safe and secure, strong and sustainable". Outside of Europe, the Melbourne Principles[6] (for sustainable cities) were drawn up in Australia during a planning session leading up to the 2002 Earth Summit held in Johannesburg, South Africa (UNEP 2002). They consist of 10 brief statements on how cities could become more sustainable, among which are the following: 1) the need to build on the characteristics of ecosystems in the development and maintenance of sustainable cities, 2) recognize and build on the distinctive characteristics of cities, 3) expand and enable cooperative networks to

5 Also known as *The Charter of European Sustainable Cities and Towns towards Sustainability*. Available at: http://www.sustainablecities.eu/the-aalborg-charter/.

6 Melbourne Principles for Sustainable Cities. UNEP, Integrative Management, Series Number I, Division of Technology, Industry and Economics. Available at: https://archive.epa.gov/bns/web/pdf/melbourneprinciples.pdf [Accessed April 10, 2021].

work towards a common and sustainable future and 4) enable continual improvement based on accountability, transparency and good governance.

Definitions of the term "sustainable city" are numerous and variable. Rogers (1998) conceptualized a sustainable city as a place where a higher quality of life is achieved, in tandem with policies that efficiently cut demand on resources outside of the city. Other scholars (e.g. Meadows 1999) approach the term from a different angle, an eco-friendly one. Their respective sustainable city representations focus on a city's ecological needs, assessing and reducing pollution and GHG emissions, as well as energy and water consumption (along with other goals). More recently, Rode and Burdett (2011) have opted for a socio-economic explanation, where social equity, together with a greener living environment, should be considered for the development of sustainable cities. They pointed out that cities should offer proximity, density and variety, which would eventually yield productivity benefits for companies, and help stimulate innovation and new job creation via high-tech clusters in big cities, for example, look at Silicon Valley.

SUSTAINABLE CITY			
Social Development	Economic Development	Environmental Management	Urban Governance
• Education and Health • Food and Nutrition • Green Housing and Buildings • Water and Sanitation • Green Public Transportation • Green Energy Access • Recreation Areas and Community Support	• Green Productive Growth • Creation of Decent Employment • Production and Distribution of Renewable Energy • Technology and Innovation	• Forest and Soil Management • Waste and Recycling Management • Air Quality Conservation • Adaption to and Mitigation of Climate Change	• Planning and Decentralization • Reduction of Inequities • Strengthening of Civil and Political Rights • Support of Local, National, Regional, and Global Links

Figure 1.1. *The four dimensions of a sustainable city (source: adapted from UN (2013)[7]). For a color version of this figure, see www.iste.co.uk/karam/general.zip*

7 United Nations (2013). World economic and social survey: Sustainable development challenges. Available at: https://www.un.org/en/development/desa/policy/wess/wess_current/wess2013/WESS2013.pdf [Accessed April 10, 2021].

Other academics (De Jong *et al.* 2015; Ahvenniemi *et al.* 2017) have recently recognized that understanding the rapports between people, their activities and the environment is key to sustainability. In this context, Bibri and Krogstie (2017) affirm that a sustainable city aims to achieve a dynamic balance between economic, environmental and socio-cultural development objectives (Figure 1.1) that are outlined by a local governance system that is characterized by increased citizen contribution.

This holistic vision of a sustainable city involves an efficient, local and balanced process that expands into all areas of local decision-making and aims to continually improve various urban systems, both design- and function-wise. The governance dimension is therefore essential as cooperative efforts from various stakeholders are needed to take an all-encompassing approach to solving the complex challenges of cities.

This rationale is hypothesized using the quadruple-helix model (Figure 1.2), in which universities, governmental authorities and the private sector join forces with the people, in order to find solutions to shared problems (Arnkil *et al.* 2010). Certainly, this archetype of *collaborative urban governance* – which is increasingly supported by ICT (Termeer and Bruinsma 2016) – has the potential to cut across outdated prerogatives and practices (Lubell 2015) and is seen as fundamental for the creation of a sustainable city (Kordas *et al.* 2015).

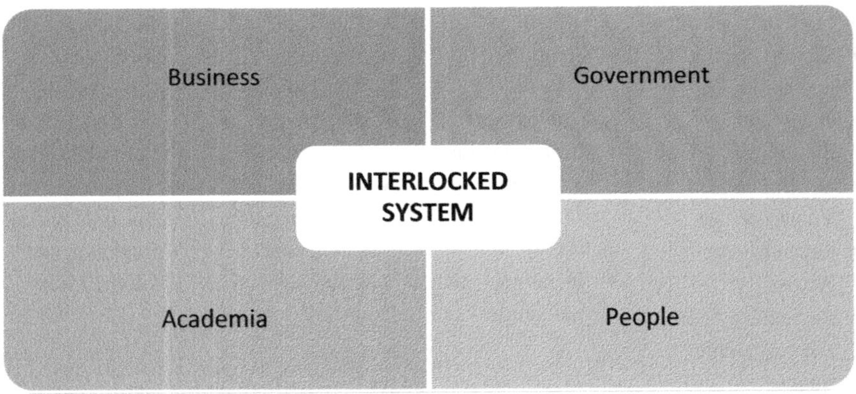

Figure 1.2. *The quadruple-helix model (source: adapted from SIP-SSC (2015)[8])*

8 SIP-SSC (2015). Strategic innovation agenda for smart sustainable cities. Available at: https://www.diva-portal.org/smash/get/diva2:812341/FULLTEXT01.pdf [Accessed April 12, 2021].

1.1.3. *The start of smart*

While the sustainable city model was traditionally seen as the best option, it has nevertheless been superseded in recent years by the smart city model (De Jong *et al.* 2015). In fact, modern understanding of a smart city is integral to the advent and growth of ICT and computing (Harrison and Donnelly 2011; Kitchin 2014). In advanced economies, strong infrastructures, cutting-edge technologies and investment in energy efficiency are all quoted as the main ICT-driven benefits conferred to cities (Bibri and Krogstie 2017).

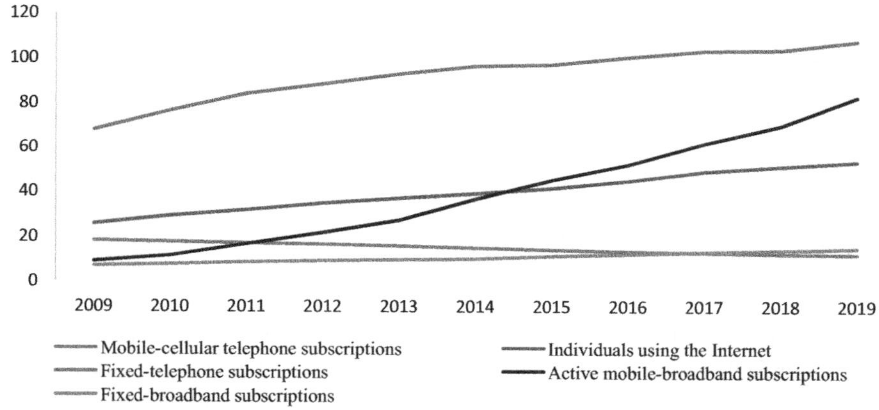

Figure 1.3. *Global ICT developments, Y2009-19 (source: adapted from ITU (2019)[9]). For a color version of this figure, see www.iste.co.uk/karam/general.zip*

Nowadays, although the fallout of the wide-scale deployment of ICT is still not fully understood – and an affordable and reliable Internet connection is still out of reach for the majority of people in the world – the network, both in terms of infrastructure and content, has grown rapidly since its inception and is stimulating vast innovation and improved user engagement (Figure 1.3). According to Turok (2014), increased deployment of ICT in the cities of emerging countries has been cited as a driver of efficacy in urban infrastructures, reducing the cost of city operations while supporting innovation and the eradication of poverty. In some instances, such as in Hong Kong and Singapore, urban economies tend to

9 ICT Development Trends and Approaches for Digital Transformation. Available at: https://www.itu.int/en/ITU-D/RegionalPresence/AsiaPacific/SiteAssets/Pages/Events/2019/RRITP2019/ASP/ICT%20Development%20Trends%20and%20Digital%20Transformation%20ITU%20ITP%202019.pdf [Accessed April 12, 2021].

circumvent the need to retrofit existing infrastructure by incorporating ICT into new ones during their initial construction stages.

From our point of view, the burden is currently also put on the shoulders of governments; they are constantly seeking to gather and monitor data in connection with governance, infrastructure, the economy and the environment (UN-Habitat 2016). Kitchin (2014) affirms that the use of data allows cities to assess their performance in various aspects (from crime prevention to energy efficiency and investment) and thus improve the quality of their public offerings. Indeed, today, performance measurement has become a key element in the eyes of both officials and planners. This is the case because the use of data enables cities to not only measure their own performance and make evidence-based decisions, but also to compare and benchmark themselves against other cities, nationally and internationally. At this point, connectivity has reached a tipping point and the IoT is gradually being replaced by the *Internet of Everything* – cleverly dubbed as *everyware* – a network of networks where billions of connections could create unparalleled opportunities in making cities manageable in new, comprehensive and dynamic ways (Greenfield 2006).

Within this context, however, Silk and Appleby (2010) have indicated that the efficient exchange of information requires data accessibility, which is only possible via the publication and standardization of basic datasets: *Open Data*. Specifically, Open Data aims to strengthen the average citizen's contribution to urban governance and is – to some extent – a commitment to transparency and accountability.

Before we conclude this present section, we wish to reiterate that the smart city concept is comparatively new and is still a contested topic and, out of interest, we proceed by asking: where did the idea of *smart city technology* come from? (Surprisingly, former USA president Bill Clinton takes all the credit for this one!) Back in 2005, through his charitable organization – The Clinton Foundation – he challenged network equipment provider Cisco to use its technical know-how to render cities more sustainable. As a result, Cisco devoted USD 25 million over a five-year period to research the topic, which led to the Connected Urban Development (CUD) program. The latter involved working with the cities of San Francisco (USA), Amsterdam (The Netherlands) and Seoul (South Korea) on a number of pilot projects to verify the technology's potential.

In 2010, the year Cisco's pledge expired, it launched its Smart and Connected Communities Division in order to commercialize the products and/or services that had been developed over the course of the program. Not long after, IBM had a similar vision to use IT (Information Technology) to make cities, say, smarter. At the time, both corporations were entirely focused on smart cities; however, their respective approaches to building those cities diverged significantly:

– IBM's strategy – on the one hand – was underpinned by its recent focus on information management and analytics. Through acquisition and internal R&D (Research and Development), the company succeeded in arming itself with analytical procedures and data processing technologies that are indispensable in understanding massive amounts of sensor data.

– Cisco's projects – on the other hand – ranged from modified or upgraded projects, such as their partnership with the Metropolitan Transit Authority in New York to improve rail and station monitoring, to completely new ones such as Songdo – a sustainable city to be built on reclaimed marshland in South Korea.

In conclusion, we wonder: did Cisco win the bet? Yes, in fact, it did because the investment made came first in the smart city make-a-thon. This instigated a global market whose current value measures up to billions of dollars, with hopeful prospects (Bloomberg's!) to grow further in the future to reach USD 253 billion by 2025[10].

1.2. Definitions, features and models

A review of academic literature reveals that the use of the term *smart city* has increased exponentially – as of early 2009 – to the extent that it became the most prevalent classification for sustainable urban development (De Jong *et al.* 2015). It was the buzzword, drawing increased attention among research agencies, universities, governments, officials and specialized companies. All over the world, there has been hasty proliferation and promotion of smart city programs and the market potential for smart city solutions is set to be substantial (Etezadzadeh 2016). China, for instance, is urbanizing quickly and the Chinese government is currently pushing for renovation throughout its cities, with over 100 initiatives in the pipeline: eco-cities, low-carbon cities, as well as smart cities (SIP-SSC 2015). Similarly, a few years ago, India announced plans to develop smart cities in response to the country's growing population and pressure on existing urban infrastructure (Smart Cities India 2015).

Today, although the concept of the smart city is still viewed favorably, implementing initiatives in countries with vastly different needs makes it difficult to establish shared definitions and common trends at the global level. That is, there is no clear-cut definition of the term *smart city*, or any general agreement on what its

10 Bloomberg, Business. November 2019 Issue. See full report at: https://reports.valuates.com/sreport/QYRE-Othe-1K199 [Accessed March 30, 2021].

inherent features are (Caragliu *et al.* 2011; Dameri 2013; Kitchin 2014; Neirotti *et al.* 2014; Albino *et al.* 2015). In Latin America, for example, smart city projects tend to be weighted towards the improvement of security, local government management and mobility, whereas in Asia, the emphasis is instead on the improvement of infrastructure and mobility; there again, in Europe the aim is to boost efficiency in terms of the public offering, establishing a socially inclusive culture and enhancing citizen well-being (Neirotti *et al.* 2014).

According to SIP-SSC (2015), a smart city uses ICT to enhance livability, workability and sustainability. Similarly, the USA Office of Scientific and Technical Information[11] affirmed that a city is qualified as smart if it integrates and monitors its critical infrastructure – including roads, bridges, tunnels, railways, seaports, water, power and major buildings – in order to optimize resources, formulate a preventive maintenance plan and monitor security while maximizing services to its citizens. Another astute smart city definition is that of Guerrero-Pérez *et al.* (2013) who suggest that a smart city could be seen as a defined geographical area capable of efficiently managing resources, as well as the waste generated by those within it. From their perspective, Georg *et al.* (2012) suggest a reasonable macro-definition by describing smart city projects as: "[...] nothing but the sum of several interconnected, smart micro-projects (e.g. infrastructure, telecom and ICT, as well as smart buildings and districts)". Thus, "a smart city project is an aggregation of a number of varied-sized and -shaped smart projects" they added.

Refer to Table 1.3 for an extensive – though not exhaustive – list of definitions for smart cities, from earliest to most recent.

Given the scope of this book and the objectives assigned to it, we devised our own definition of a smart city, which is as follows: "A smart city – we believe – is a city that is conceived, developed and run by the government and the people, both being smart, for the sake of promoting citizens' well-being and ensuring a better quality of life for all".

The term *smart city* first emerged out of the increased use of ICT, big data, the IoT and ubiquitous computing in cities. According to Caragliu *et al.* (2011), the smart city concept was described as a strategic tool to encompass modern urban production factors in a common framework and, in particular, to highlight the importance of ICT for enhancing the competitive profile of a city. So, to recap, a smart city is one that uses ICT strategically in its administration to provide efficient services to citizens, monitor policy outcomes, manage and optimize existing infrastructure, and employ cross-sector collaboration and enable new business models.

---

11 Found in Eremia *et al.* (2017).

Definition	Author/s, Year
A smart city is one that monitors and integrates the conditions of all of its critical infrastructures to better optimize its resources, plan its preventative maintenance activities, and monitor security aspects while maximizing services to its citizens.	Hall (2000)
A smart city performs in a forward-looking way in terms of its economy, people, governance, mobility, environment and living, and is built on a combination of endowments and the activities of self-decisive, independent, and aware citizens.	Giffinger *et al.* (2007)
A smart city connects its physical infrastructure, ICT infrastructure, social infrastructure and business infrastructure to leverage the collective intelligence of the city.	Harrison *et al.* (2010)
Smart cities use smart computing technologies to make the critical infrastructure components and services of a city – which include city administration, education, healthcare, public safety, real estate, transportation and utilities – more intelligent, interconnected and efficient.	Washburn *et al.* (2010)
A city is smart when investments, in human and social capital and traditional and modern communication infrastructure, fuel sustainable economic growth and a high quality of life, with a wise management of natural resources, through participatory governance.	Caragliu *et al.* (2011)
A smart city infuses information into its physical infrastructure to improve convenience, facilitate mobility, add efficiency, conserve energy, improve the quality of air and water, identify problems and fix them quickly, recover rapidly from disasters, collect data to make better decisions, deploy resources effectively and share data to enable collaboration across entities and domains.	Nam & Pardo (2011)
Smart cities should embody two main ideas: 1) Smart cities are all about networks of sensors, smart devices, real-time data and ICT integration in every aspect of human life; and 2) Smart cities should do everything related to governance and economy using new thinking paradigms.	Cretu (2012)
A smart city is a high-tech intensive and advanced city that connects people, information and city elements using new technologies in order to create a sustainable, greener city, competitive and innovative commerce and increased quality of life.	Bakici *et al.* (2012)
A smart city is a well-defined geographical area, in which advanced technologies cooperate to create benefits for citizens in terms of well-being, inclusion and participation, environmental quality, intelligent development; it is governed by a well-defined pool of subjects, able to state the rules and policies for the city government and steer development.	Dameri (2013)
Smart cities use data, information and IT to provide more efficient services to citizens, to monitor and optimize existing infrastructure, to increase collaboration among different economic actors, and to encourage innovative business models in both the private and public sectors.	Marsal-Llacuna *et al.* (2015)
The smart-city concept is a tool for the creation of sustainable communities. It incorporates a mixture of transparent and collaborative governance schemes where citizen participation is key.	Simard (2015)

Table 1.3. *Defining smart cities (source: compiled from various sources)*

Schaffers *et al.* (2011) similarly identify three key areas for potential smart city applications: the innovation economy, city infrastructure and utilities and city governance. They argue that in order to become smart, a city needs to (1) create a rich environment of broadband networks that support digital applications; and (2) initiate large-scale participatory innovation processes for the creation of applications.

The smartness of a city is thus determined by a set of physical and legislative infrastructures (Eremia *et al.* 2017) that support economic development, ensure social inclusion and enable environmental protection (Figure 1.4). Table 1.4 highlights the direction of developments in a city and provides examples of various applications through which the main objectives of a smart city could be achieved.

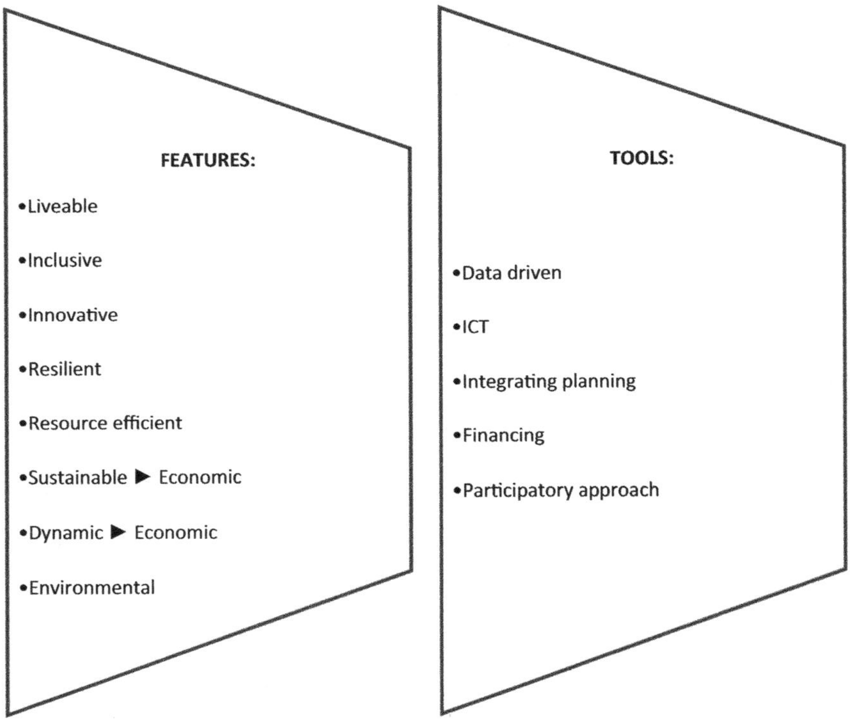

Figure 1.4. *Smart cities' features and tools (source: adapted from Eremia* et al. *(2017))*

We now progress by revealing the main smart city models, explicitly those of Giffinger *et al.* (2007), Cohen (2012) and Tobergate and Curtis (2014). The holistic models of Giffinger *et al.* (2007) and Cohen (2012) are largely seen to be the most illustrative in their capacity to demonstrate and exhibit the strategic components of a smart city by grouping them into six different areas (Figure 1.5).

Area of application	Description	Examples
Smart Buildings	Incorporating the advantages of communication and control systems.	Optimizing the heating systems, ventilation and air conditioning.
Education, Health	Apps that improve the activity in these domains and ensure all citizens have access to high-quality services.	Monitoring systems of the elderly, monitoring by telemedicine.
Smart Energy	Smart electrical energy system that interconnects all utilities and end-users via smart infrastructures.	Smart Grid apps, optimization of network operation, complying with environmental standards, smart lighting.
Smart Grid	Real-time consumption metering of energy, water and natural gas.	Online information about consumption, wireless smart meters.
Smart Utilities	Intelligent management of the water distribution system and wastewater.	Smart wastewater systems, real-time solid waste monitoring.
Smart Parking	Managing parking places using sensors and CCTV.	Monitoring systems of the vehicles.
Integrated Supply Systems	Synchronizing supply and demand; measurement, monitoring and organization of the transportation around the supply chains of the cities.	n/a
Smart, Integrated Transport	Traffic monitoring and real-time optimization using all transportation means.	CCTV for traffic, smart parking networks, minimizing the impact on the environment.

Table 1.4. *Direction of developments in smart cities (source: adapted from Meijer* et al. *(2016))*

According to Giffinger *et al.* (2007), smart cities should be established on the basis of a smart combination of funding and activities, exactly as suggested by connected, self-governing intelligent citizens. Table 1.5 offers in-depth insights into the main components of a smart city.

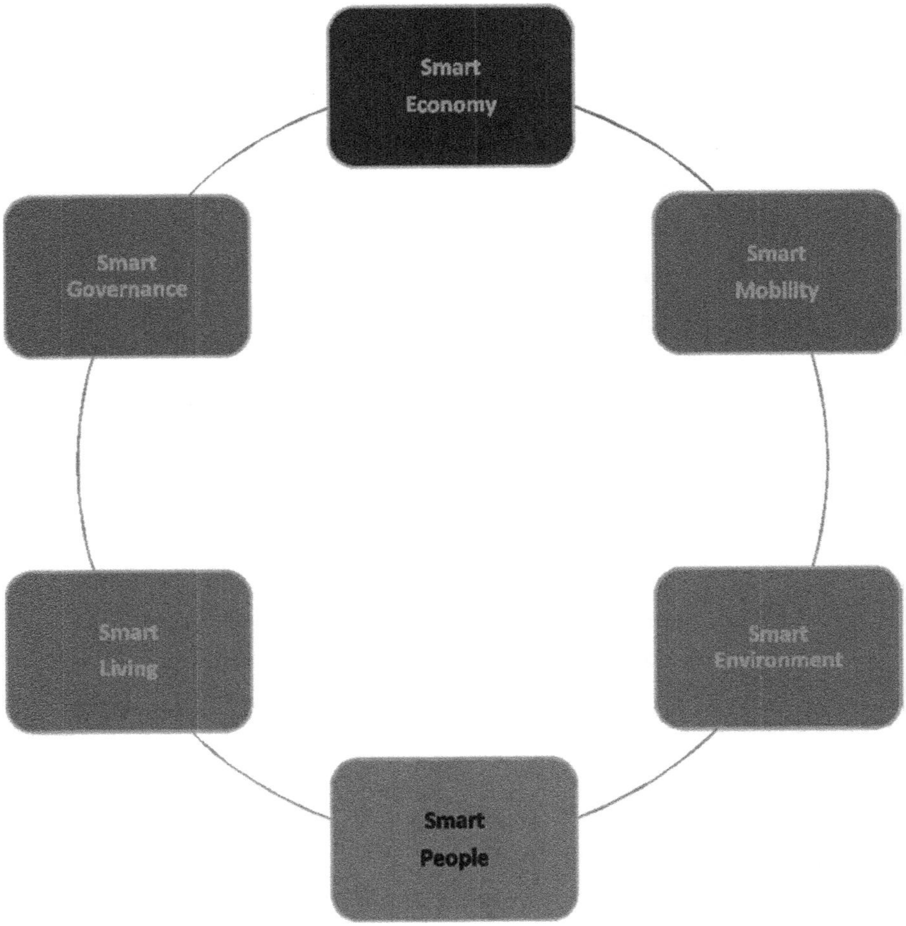

Figure 1.5. *The main components of smart cities (source: adapted from Giffinger* et al. *(2007))*

These components were established in 2007 – as part of the European Smart Cities initiative – and therefore numerous authors (Batty *et al.* 2012; Lazaroiu and Roscia 2012) have referred back to them year after year to develop more elaborate frameworks and strategies. Boyd Cohen's (2012) Smart City Wheel is one example (see Figure 1.6).

▼Smart Economy:

- Innovative spirit
- Entrepreneurship
- Economic image, trademarks
- Productivity
- Flexibility oflabor market
- International embeddedness
- Adaptation

▼Smart People:

- Level of qualification
- Affinity to life-long learning
- Social and ethnic plurality
- Flexibility
- Creativity
- Cosmopolitanism, open-mindedness
- Participation in public life

▼Smart Governance:

- Participation in decision-making
- Public and social services
- Transparent governance
- Political strategies and perspectives

▼Smart Mobility:

- Local accessibility
- National and international accessibility
- Availability of ICT infrastructure
- Sustainable, innovative, and safe transport systems

▼Smart Environment:

- Attractiveness of natural conditions
- Environmental protection
- Sustainable resource management

▼Smart Living:

- Cultural facilities
- Health conditions
- Individual safety
- Housing quality
- Education facilities
- Touristic attractivity
- Social cohesion

Table 1.5. *A deep dive into the main components of smart cities (source: adapted from Giffinger* et al. *(2007))*

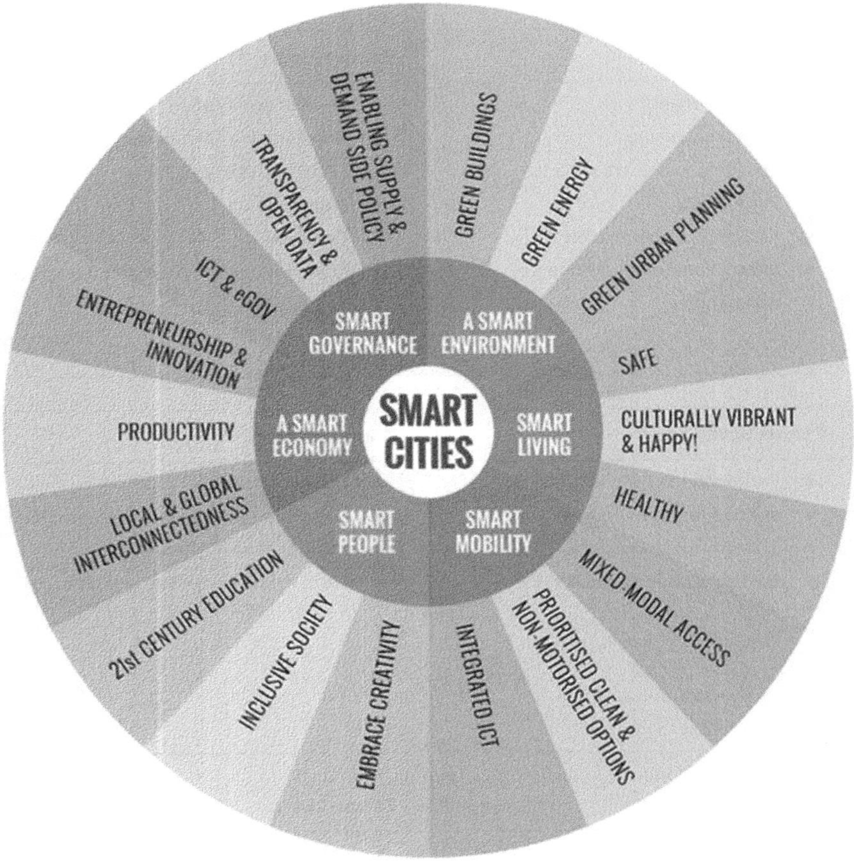

Figure 1.6. *Smart City Wheel (source: Cohen (2012)). For a color version of this figure, see www.iste.co.uk/karam/general.zip*

A synthesis of the works of Giffinger *et al.* (2007) and Cohen (2012) by Simard (2015) allows for a better understanding of the relevance of smart city components vis-à-vis teething problems encountered by urban settlements today. According to Simard, it is crucial to have the support of governing bodies (as a first step) *ex ante* the start of any project process. Further, she states that, within the realm of smart cities, citizens must necessarily take part in this process by becoming active stakeholders and thus contributing to its success. To build on the work of Simard (2015), we chose to classify smart city components as primary or secondary. In our view, once the primary components have been securely rooted and established, the other (secondary) ones – four in total, all of equal importance – should follow (Figure 1.7). Indeed, no matter how well-governed a city is or how many smart people are part of it, we maintain that it cannot be qualified as smart unless it can demonstrate that all six primary and secondary components are in place.

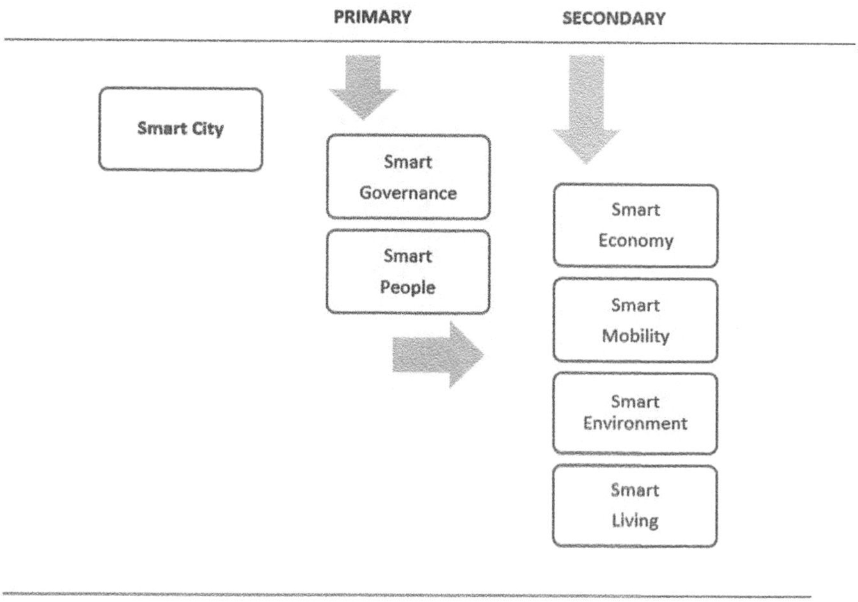

Figure 1.7. *Giffinger* et al.*'s (2007) work revisited (source: created by the author – inspired by Simard (2015)). For a color version of this figure, see www.iste.co.uk/karam/general.zip*

Tobergate and Curtis (2014) have also made great use of existing work on smart cities by proposing a coherent illustration of the different aspects of a smart city. More precisely, they have added three factors (or rather, *headings*), under which all six smart city dimensions can sit.

Technology Factors	Human Factors	Institutional Factors
• Physical infrastructure • Smart technologies • Mobile technologies • Virtual technologies • Digital networks	• Human infrastructure • Social capital	• Governance • Policy • Regulations and directives

Table 1.6. *Core factors of smart cities (source: adapted from Tobergate and Curtis (2014))*

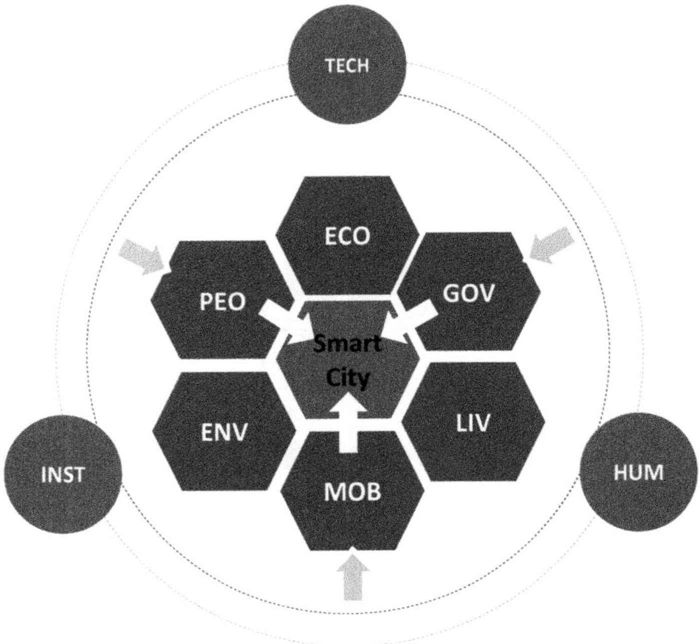

Factors	Components
TECH: Technology	ECO: Economy
INST: Institutional	PEO: People
HUM: Human	ENV: Environment
	MOB: Mobility
	LIV: Livability
	GOV: Government

Figure 1.8. *Smart city: factors versus components (source: adapted from Tobergate and Curtis (2014)). For a color version of this figure, see www.iste.co.uk/karam/general.zip*

As reported by Tobergate and Curtis, the factors (first layer) can be broken down to get to the components (second layer) (Figure 1.8); the factors, through their respective parts (see Table 1.6), would then act as a conductive thread for components (primary and secondary) that, combined, would ultimately fuel the development process of smart city strategies around the world. The next section features some real-life, global examples of smart cities and initiatives. Using easy-to-understand language, it briefly exposes international developments and improvements made in this field.

1.3. Examples of smart cities and initiatives

From smarter transit systems to greener buildings, smart cities are quickly transforming the way people think about urban living. To elucidate this global penchant for smarter urban living, five examples of smart projects and initiatives from around the world are given below. The proposed smart city tour begins in Europe (Spain) and ends in Asia (South Korea); the fifth edition of the IESE Cities in Motion Index (CIMI) inspired our choice of smart cities[12].

As mentioned, the idea of *smart cities* is vague, with myriad associations (smart living, smart security, smart governance, smart people, smart anything and everything), and its connotations vary across continents, nations and cities (even within the same nation). We believe this reasoning helps us to substantiate the subjective nature of the term. Furthermore, though smart cities are relatively abundant today, history has shown that not all of them were destined to flourish (*Remember?* A top-down approach to building a smart city is likely to fail, particularly given that stakeholders often misinterpret the term *smart city*).

However, we must state here and now that not all smart cities are equally successful, genuine or human; thus, we raise the red flag on Songdo (South Korea) and Dubai (United Arab Emirates) for being mere *technology demonstrators* rather than *habitable cities* (Rochet 2014).

1.3.1. *Barcelona, Spain*

With 22 programs encompassing over 120 active projects that incorporate all areas of city management – from public space and mobility to open government – Barcelona was named European Capital of Innovation 2014[13]. A year later, in 2015, it was labeled – along with New York, London, Nice and Singapore – one of the world's smartest cities.

Since 2011, the city has worked hard and fast to turn the hypothetical smart city concept into a reality. Evidently, they have succeeded in doing so! Investment-wise,

12 IESE (2019). Cities in Motion Index 2019. Available at: https://blog.iese.edu/cities-challenges-and-management/2019/05/10/iese-cities-in-motion-index-2019/ [Accessed March 15, 2021].

13 Data in section 1.3.1 was taken from a consultancy report by Rahyaputra V., Khawarizmy N.M. and Saputri N.R., entitled "Barcelona's smart city: The frontrunner in digital transformation". Available at: http://cfds.fisipol.ugm.ac.id/uploads/files/posts/72/CFDS_CASESTUDIES_BARCELONA.pdf [Accessed December 27, 2019].

Cisco turned out to be one of the city's key technology providers. It invested almost USD 30 million to build a new center focusing on the IoT for urban systems. Cellnex Telecom, a wireless telecoms company, then integrated existing city services into one system (the Urban Platform of Barcelona) and has been running the city-wide network ever since. Due to its extensive use of the IoT and the successful implementation of its strategies, Barcelona is now regarded as one of the global leaders in the smart city market. Actions implemented under the Barcelona Smart City strategy lie on three axes: international promotion, international collaboration and local projects. Among the various smart projects Barcelona has developed in recent years, we cite the following:

– the Apps4 BCN portal, an application that provides a virtual meeting point portal for people to look for apps and improve their experience of the city;

– smart transportation through which the city aims at showing its commitments in becoming the frontrunner in the utilization of renewable energy, particularly in the transit sector (e.g. the development of electric vehicles);

– smart education, the so-called Smart City Campus, enabling supportive environments for businesses, research centers and universities to promote synergies and generate an urban laboratory;

– smart governance via the establishment of an e-government program, the Open Data BCN, though which Barcelona intends to open up the city database to the public, in order to increase transparency and credibility; and also to provide and universalize open data access to reach all stakeholders. Beyond this, Barcelona has created a Smart City Personal Management Office with the purpose of coordinating and synchronizing all city projects and operations deemed *smart*.

1.3.2. *Amsterdam, The Netherlands*

Ordinarily, smart city projects should benefit the people who live, work and visit the city in which they are implemented[14]. In our view, these projects are a dynamic way of using technology to build a better working world. Amsterdam's smart city initiative has now progressed from the early stages of experimentation, a trial-and-error phase which culminated in the city opening its data up to the public. Now in its second stage, the initiative is characterized by more solicitous and

14 Inspired by a consultancy report by Frank Harmsen (E&Y Advisory) entitled "Amsterdam's Intelligent Approach to the Smart City Initiative". Available at: http://marketing.mitsmr.com/PDF/EY-SponsoredViewpoint-Amsterdam.pdf [Accessed March 15, 2021].

focused efforts that will ultimately enable the city to spur innovation and improve the quality of its offering. Table 1.7 below shows the main lessons that can be learned from Amsterdam's experience in the area of smart city development.

No.	Strength	No.	Strength
1	Strong governance: • The backing of the political establishment is critical!	2	Proof-of-concept projects: • Pilot projects are key! • Data matter, whether big or small.
3	Human-centric approach: • Listen to citizens' needs and propose suitable solutions.	4	Create a pool of talents: • Attract and retain talented people. • Develop university programs dedicated to smart cities.
5	Build strategic alliances: • Work with partners, inside the city government and outside.		

Table 1.7. *The five strengths of Amsterdam's smart city (source: Ernst and Young). For a color version of this table, see www.iste.co.uk/karam/general.zip*

1.3.3. *London, United Kingdom*

London is a smart city global leader, known for developing innovative solutions and tackling urban challenges by making data accessible to all, at all levels[15]. Indeed, this was possible thanks to the London Datastore, a free and open data-sharing portal with over 700 datasets, launched back in 2010, and also to the recently implemented London Office of Data Analytics (LODA) program, which allows for increased data sharing and collaboration for the benefit of all Londoners. The *Smarter London Together* roadmap was first set in 2013 – and last updated in 2016, suggesting a new approach to building a smarter London based on collaborative activity between the city's local authorities and public services. The city's smart strategies incorporate plans to (among others): upgrade transit systems, improve the environment, shrink health inequalities, render housing affordable to all and stimulate economic development (Table 1.8).

15 Section 1.3.3 summarizes the London Mayor's roadmap to transform London into the smartest city in the world. The report is entitled "Smarter London Together". Copies of the report are available at: www.london.gov.uk/smart-london [Accessed March 15, 2021].

Assignments	Strengths	Best practices
• More user-designed services • Strike new deals for city data • World-class connectivity and smarter streets • Enhance digital leadership and skills • Improve city-wide collaboration	• City Data (London Datastore) • Data, tech and the environment (CleanTech, FlexLondon) • Data, tech and transport (Transport for London TfL)'s track record) • Data, tech and safety (Public Interactive Dashboards) • Data and test-beds (Queen Elizabeth Olympic Park)	• Put people first and respect diversity when designing digital services or adopting new technology • Lead in data innovation but build trust and transparency in how public data is used • Be better connected and open to new tech in the built environment • Strengthen digital leadership in public services and enhance the digital skills and understanding of citizens • Make city-wide collaboration and tech partnerships better to design and share what works for citizens across public and community services

Table 1.8. *Towards a smarter London (source: Greater London Authority, 2018: www.london.gov.uk/smart-london)*

1.3.4. *Copenhagen, Denmark*

Copenhagen[16], in contrast to other comparable cities around the world, is focusing its efforts on promoting the use of green technology and becoming carbon neutral by the end of 2025 (Figure 1.9). Today, the city is at the heart of clean technology innovation – it is Europe's cleanest (and healthiest) capital – and is one of the greenest capitals in the world. Within this framework, the city has succeeded in reducing its carbon emissions by over 50% (between 1996 and 2025). Huge investments have been made in this respect, totaling USD 1 billion.

Bicycle lanes and super cycle highways were created, enabling half of the city's residents to cycle (rather than drive) to work or school; Copenhagen's vision is to become the world's most bicycle-friendly city. Moreover, in order to alleviate congestion on the roads, the city installed a smart traffic management system to optimize traffic flow; it also encouraged citizens to adopt green modes of transport by setting a dynamic pricing model in motion that consists of adjusting toll and

16 Data taken from a presentation prepared by local authorities, entitled: "Copenhagen Smart City". Available at: http://e97f7d10b0a403e208e5-9fbee7de8d51db511b5de86d75069107.r75.cf1.rackcdn.com/Copenhagen.pdf [Accessed December 27, 2019].

parking rates on demand. Aside from that, Copenhagen has gone further by installing cost-efficient waste collection and recycling systems, allowing the city to recycle half of its waste, most of which is used to fuel its district heating network (to which approximately 98% of households are connected). The city has also installed an energy-efficient district cooling system that relies on extracted cold seawater and leads to a substantial reduction in energy consumption – nearly 70%, relative to a conventional air-conditioning system. (see Table 1.9 for a list of some of the elements of smart city projects and initiatives in Copenhagen).

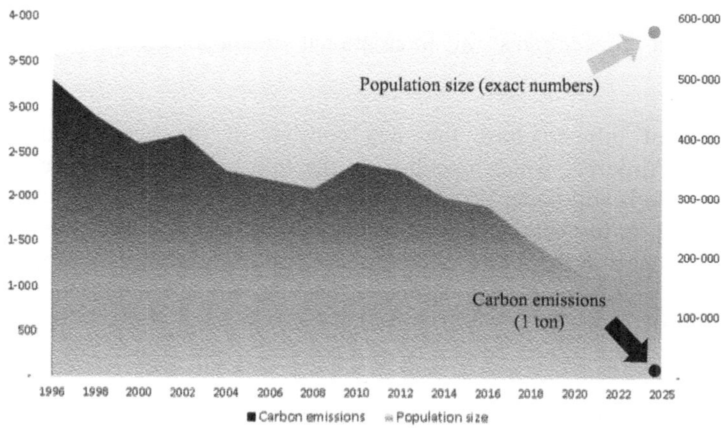

Figure 1.9. *Copenhagen: carbon emission versus population size (source: compiled from various sources). For a color version of this figure, see www.iste.co.uk/karam/general.zip*

Smart Projects:	**Smart City Initiatives:**
Work with: — targeted use of data in solving problems — new technology or known technology in new ways — efficient use of the city's resources — new ways of involving citizens or companies	• Cooperation across administrations • Project coordination board ▶ One "*strategi*" for smart city • Focus on lighthouse projects: — Open city data platform — City map — Big data platform with partners — City solution lab • Mobility projects and ITS (Intelligent Transport Systems) • Digital infrastructure

Table 1.9. *Copenhagen: smart projects and initiatives (source: compiled from various sources). For a color version of this table, see www.iste.co.uk/karam/general.zip*

1.3.5. *Seoul, South Korea*

Songdo international city is part of the Incheon Free Economic Zone, along with Yeongjong and Cheongna. The function of each region is as follows (Sang Keon *et al.* 2016)[17]:

– Songdo is in charge of international business relating to IT and BT (Business Technology), as well as R&D;

– Yeongjong focuses on aviation logistics, tourism, and leisure; and

– Cheongna concentrates on international finance, high-tech industry and distribution.

In terms of infrastructure, the area consists of the very famous Incheon International Airport[18], which operates as the hub airport of northeast Asia. At present, Songdo is still under development, with plans to invest a few extra billions in it until 2022, to furnish the city and strengthen its infrastructure to accommodate additional residents in the future. Moreover, as part of the country's project to build a cutting-edge green city, Songdo has been converted into a U-city (Ubiquitous-city) (see Table 1.10).

• U-city construction and basis • Public-private partnership corporation	• Construction work • Integrated Operation Center	• Spread public-private partnership service model • Enlarge overseas export
2005-2014	2015-	2020-2022
U-City Construction Plan — Introduction period	***U-City Construction*** — Construction period	***City Competitiveness*** — Spread period

Table 1.10. *Establishing Songdo city, targets and steps (source: adapted from Sang Keon* et al. *(2016))*

In general, the U-city in question is divided into public and private services (Figure 1.11). More precisely, public services are designed to provide around-the-clock services in various categories (traffic management and crime and

17 "International Case Studies of Smart Cities: Songdo, Republic of Korea". Report published June 2016. Available at: https://publications.iadb.org/en/international-case-studies-smart-cities-songdo-republic-korea [Accessed March 15, 2021].

18 Ranked first for airport services (for 10 consecutive years), and second for international cargo transportation.

disaster prevention)[19], while private services are intended to provide facilities that will improve the quality of people's day-to-day lives. As for eco-friendly services (U-bike, U-street, U-foreigner Support Mobile Service, etc.), they are offered as specialized services through the city's U-IT platform.

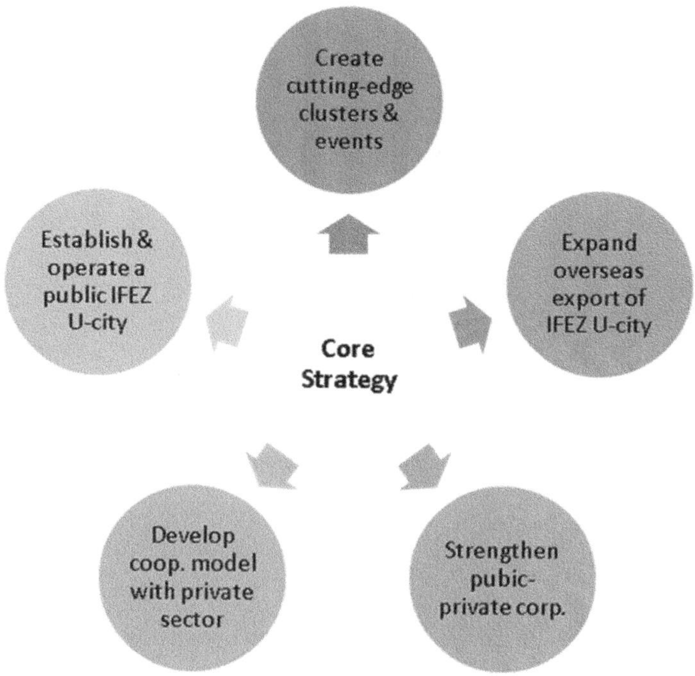

Figure 1.10. *Songdo U-city's core strategies (source: adapted from Sang Keon* et al. *(2016))*

Songdo U-city was based on a top-down managerial approach and executed using comprehensive plans and construction procedures. Tested in 2014, these procedures were used two years later, in 2016. However, the U-city looks like a front-line creation today; (sadly!) its present-day structure requires quite a few alterations *ex ante* the implementation of services. We believe that issues arose because of significant differences in what was requested and what was required at both the conception and construction stages of the project. In fact, technologies have been rapidly developed (in the past decade at least!) and citizens' request levels

19 Songdo U-city collects 24-hour real-time data from on-site equipment such as CCTV, sensor devices and traffic detectors. Systems in Songdo are being built so that the data gathered could eventually be stored and analyzed in order to provide citizens, tourists and businesses with expedient public offerings.

(standards) have risen too. Therefore, aside from the high probability that the technology implemented will already be outdated upon project completion, the plans of related entities and on-site equipment need to be clearly defined as well, to avoid wasting the budget on duplicate systems, for example.

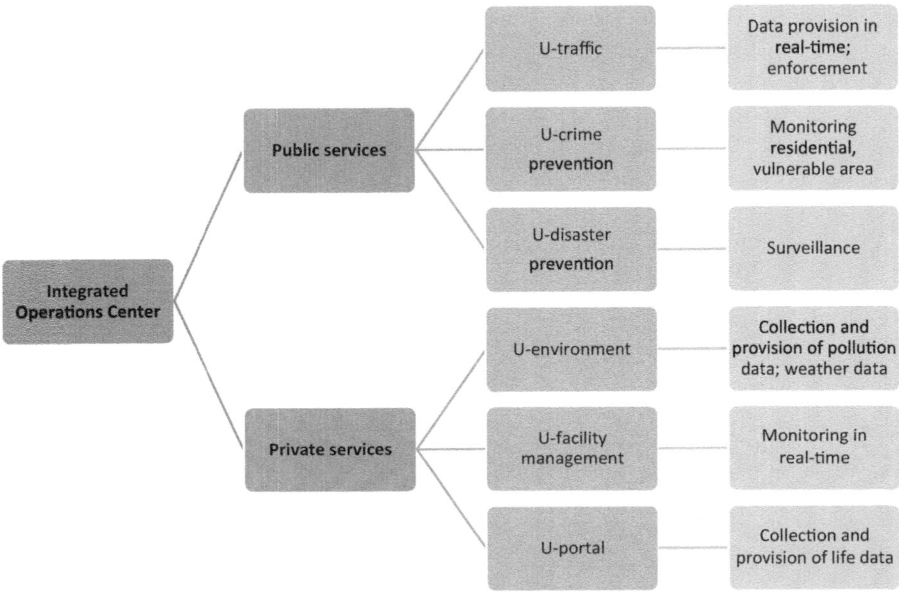

Figure 1.11. *Songdo U-city integrated operations center (source: adapted from Sang Keon* et al. *(2016))*

Before moving on to the next section, in which an overview of the size and potential of the global smart city market is presented, we close this one by sketching a comparative table that summarizes the main components of 15 different smart cities.

Table 1.11 shows the various similarities and differences between international smart city models. If the data is interpreted column by column (i.e. read top to bottom), then the scores can be compared and the frontrunners invited up to the podium. However, if it is considered row by row (i.e. read left to right), the components could be ranked in order of importance[20].

20 The data shown in Table 1.11 is only indicative and must not, in any way, be seen as entirely representative. It was compiled for the purpose of this research project only and may include errors and/or inaccuracies.

SMART COMPONENTS ▼	1	2	3	4	5	6	7	8	9	10	11	12	13	14	15	TOTAL (15 PTS.)
A. MOBILITY																
— Traffic Management		•	•	•	•	•	•	•	•	•	•	•		•	•	13
— Smart Infrastructure, Roads, Lighting	•	•	•	•				•	•	•	•	•			•	10
B. SOCIAL, LIVABILITY																
— Community Work, Service	•	•	•		•	•		•	•	•	•	•	•		•	12
— Safety, Security: Crime Prevention		•		•	•	•			•	•	•	•		•		9
— Safety, Security: Disaster Prevention		•				•						•		•		4
— Street Cleanliness	•	•		•				•	•							5
— Facility Management		•						•	•							3
C. ECONOMY																
— Trade Patterns (Import, Export)																-
— Job Creation, Opportunity		•								•			•		•	4
— Technology, IoT, Big Data	•	•	•	•	•	•		•	•	•	•	•	•	•	•	14
— Inclusive Growth		•		•					•				•			4
D. GOVERNANCE																
— Open Data, Transparency	•	•	•		•	•		•	•			•	•		•	10
— City-Wide Collaboration, PPP	•	•	•	•	•	•		•	•	•		•	•		•	12
— Managerial: Bottom-up Approach		•		•	•			•					•		•	6
— Human-Centricity, User-Designed Service		•	•	•	•			•	•				•			7
— One *Strategi*				•					•							2
E. PEOPLE																
— Education, Talent		•	•		•			•			•		•		•	7
— Improved Connectivity: In-In, In-Out	•	•	•	•	•	•		•	•	•		•	•		•	12
F. ENVIRONMENT, HEALTH																
— Less Pollution, Carbon Emission	•	•	•	•	•		•	•	•	•	•				•	11
— Waste Management, Recycling	•	•		•	•		•	•	•	•	•	•			•	11
— Green Modes of Transport, Buildings	•	•	•	•		•	•	•	•	•	•				•	11
— Healthcare						•				•	•	•				4
— Usage of Renewable Energy		•	•	•	•		•	•	•	•	•				•	10
G. **OTHER : ADAPTABILITY, FLEXIBILITY**	•	•	•	•	•			•	•	•	•		•		•	11
TOTAL (24 PTS.)	11	21	13	16	14	10	5	17	18	14	12	11	11	4	15	

LEGEND:
1= Barcelona, Spain; 2= Amsterdam, The Netherlands; 3= London, UK; 4= Copenhagen, Denmark; 5= Seoul, South Korea; 6= Singapore, Singapore; 7= San Francisco, USA; 8= Helsinki, Finland; 9= Dubaï, UAE; 10= Nice, France; 11= Oslo, Norway; 12= Shanghai, China; 13= Milan, Italy; 14= Lahore, Pakistan; 15= New York, USA

• indicates ***main or primary components*** of respective smart city models. Unmarked components are relatively less significant.

Table 1.11. *Components of international smart city models: comparison (source: created by the author)*

1.4. Smart city market outlook

Big cities now face growing pressure due to the snowball effect of urbanization. Therefore, the need for operative solutions for managing urban settlements and improving sustainability and livability has eventually led to the global surge in smart cities (Chui *et al.* 2018). Principally, smart cities denote the incorporation of ICT to boost the performance and quality of urban offerings in order to reduce resource consumption and costs[21]. Indeed, global ICT companies[22] have been steadily pushing for the adoption of their technologies – a major catalyst in industry growth. Nonetheless, smart cities are developed and applied across a huge number of different domains and the smart city market, comprising of myriad sectors, is in no way exclusive to ICT. That is, all companies (regardless of the sectors they operate in) will eventually be forced to reconfigure their value chains and adapt their business models to up-and-coming market changes.

Driven by renewed interest from the industry, availability of technology and full participation of industry stakeholders, the global smart city market is expected to reach USD 2.3 trillion by 2023 and grow at a continuous CAGR (Compound Annual Growth Rate) of 29% thereafter[23]. Furthermore, estimates by Frost and Sullivan[24] suggest that investments to be made in the smart city market, covering smart solutions that would lead to the adoption of a specific mix of smart parameters by cities, may possibly reach USD 1.6 trillion by 2020. Concisely, the global smart city market is segmented based either on location (Figure 1.12): North America, Europe, Asia-Pacific, Latin America and the Middle East and Africa; or on application (domain): building, healthcare, transportation, infrastructure, governance, security and energy (Figure 1.13).

21 The weakness of the current economic model lies in the fact that it is not sustainable, in the sense that it would reproduce the same negative externalities or dysfunctionalities encountered today (carbon emissions, waste production and urban nuisances).

22 ABB Limited, Accenture, Ericsson, Oracle Corporation, Hitachi, IBM, Schneider Electric, Cisco, Microsoft, Intel, Honeywell International, Huawei and Siemens (among others).

23 Transparency Market Research (2014). Rising trend of automation to drive global smart cities market growth. Available at: https://www.transparencymarketresearch.com/pressrelease/smart-cities-market.htm [Accessed April 19, 2021].

24 Smart Cities: Frost and Sullivan Value Proposition. Available at: https://ww2.frost.com/wp-content/uploads/2019/01/SmartCities.pdf [Accessed April 19, 2021].

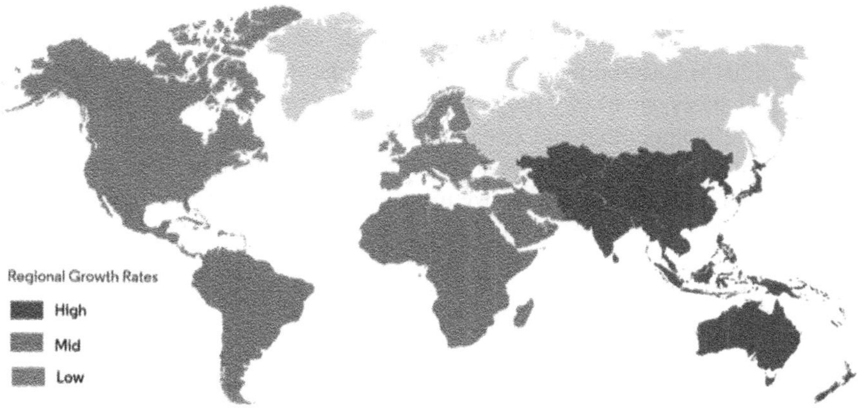

Figure 1.12. *Smart city market: growth rate by region, 2019–2024 (source: Mordor Intelligence[25]). For a color version of this figure, see www.iste.co.uk/karam/general.zip*

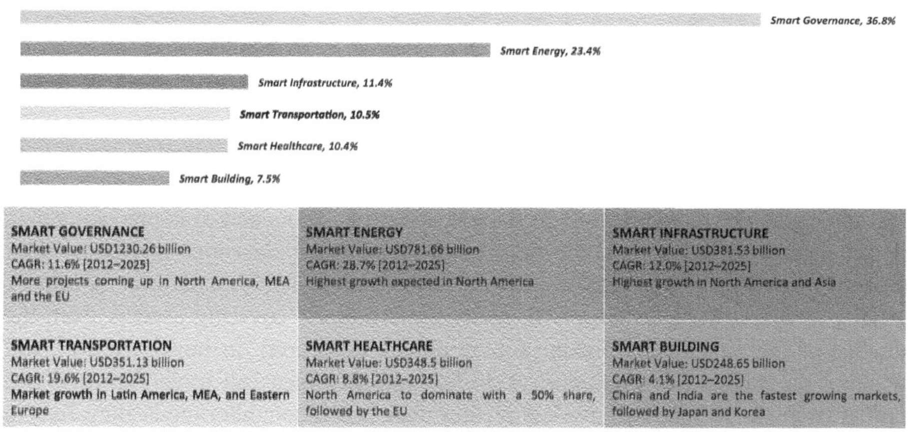

Figure 1.13. *Global smart city market, by segments: 2016 (source: Frost and Sullivan Analysis[26]). For a color version of this figure, see www.iste.co.uk/karam/general.zip*

Today, the smart city market stands to benefit from the initiative taken by governments, residents and businesses to implement projects and solve (pending) urban problems (Townsend 2013). The *Smart Cities Council*[27] is a great example of

25 Smart Cities Market – Growth, Trends and Forecast. Available at: https://www.mordorintelligence.com/industry-reports/smart-cities-market [Accessed April 19, 2021].

26 Smart Cities: Frost and Sullivan Value Proposition. Available at: https://ww2.frost.com/wp-content/uploads/2019/01/SmartCities.pdf [Accessed April 19, 2021].

27 For more information, please visit: https://smartcitiescouncil.com/.

a business-led initiative – the sector's oldest and largest smart city network – that is sponsoring the interests of its members through advocacy and lobbying actions (SIP-SSC 2015). Certainly, for city officials, national governments and supranational states, smart city projects offer the enticing potential of creating more livable, secure, competitive and sustainable settlements – the reason why they are presently on the lookout for greater-than-before deregulation and privatization (Kitchin 2014). Within this context, we move on by asking: *Are all cities destined to be smart*?

1.5. Are all cities destined to be smart?

Through their studies of several European cities, Caragliu and Del Bo (2016) acknowledged that every city in the world has the potential to become smart. According to the authors, the smart city model is one of urban developments, a go-smart path along which cities could be spotted at various points. Chui *et al.* (2018), on another note, stated that *smartness* is an ongoing process and that lots of work is still to be achieved, even by the most advanced cities, to be able to cross the smart city make-a-thon finish line. "*In fine*, a smart city should be labeled by the objectives it intends to achieve rather than the countless solutions it has to offer – and so, a smart city is not a goal by itself, but a means to an end" they added.

By and large, smart cities cannot be reduced to the mechanical solutions that are implemented within them in order to rationalize operations. This is the case because this idea goes above and beyond to cover a series of factors and/or components that, if put together efficiently, could help cities make better decisions and boost their citizens' quality of life (Rochet 2014; Etezadzadeh 2016). Literarily speaking, *quality of life* is a vague term with many dimensions – covering aspects from the quality of the air citizens breathe (environment, pollution), to how safe they feel walking the streets (safety, security), to how quickly then can travel between two points (transit, traffic), to how happy they feel in the neighborhoods they live in (sociability, livability), to a whole array of different things (Chui *et al.* 2018). This, to some extent, infers that *smart city* and *quality of life* are two inter-related notions, one leading to the other, both having the same resolves and goals.

Bettencourt *et al.* (2007) and Harrison and Donelly (2011) insist on the fact that it is better to regard a city as an *adaptive complex system* – a *system of systems*. Therefore, a given city – a complex living ecosystem of social interactions, that is – cannot be deemed smart unless the synergies and connections between its citizens are also considered to be smart. This idea was originally endorsed by Dedijer and Jéquier (1987) and Cronin and Davenport (1991) who thought *citizens* to be at the heart of smart cities – not as mere receivers of information, but as producers too. At present, after years of trial-and-error, policy-makers around the world have finally realized that smart city projects start and end with people (Chui *et al.* 2018).

Furthermore, in order to circumvent dysfunctional urbanization (which was a popular sensation in the two past centuries!) – a smart city, in general, cannot be created if it is disconnected from its territorial history, culture, tradition and evolution (Etezadzadeh 2016). However, it should have its own modus operandi, and be conceived and built based on nation-specific forms. (*Remember*? There is no common model for smart city development – and a city cannot settle for copying good practices from other cities by importing and applying them *ex nihilo.*) Here and now, however, the smart city development phenomenon seems to have become viral, a global trend – its spread was (and it still is!) disproportionate – and here is why:

– consortiums held to support the smart city market have only (at least so far) focused on specific regions and domains;

– an absence of a systemic model that enables the global adoption of ICT as well as cohesive smart cities' visions and strategies;

– growing concerns with regards to data privacy and security;

– a pronounced need among budget-conscious cities to verify ROI (Return on Investment) prior to a wider execution of smart plans;

– lack of multi-stakeholder buy-in – and adequate financial models and resources that would support the development of smart city markets.

And so, we answer: by competently using their own resources and endowments, refining communication infrastructure and web-based services, allotting ample financial resources and adopting appropriate funding schemes and creating apposite value chains that ensure a fair interplay between stakeholders, all cities around the world could become smart, at different rates, each in its own form.

1.6. Conclusion

Up until now, we have shown that *connectivity* is key – and that *functional geography* has taken over. This new reality has reshaped and wired connections between countries; it has helped ease political tensions, mitigate environmental risks and uphold economic growth. Furthermore, we have indicated that people tend to migrate to big cities in search for more connectivity and opportunities. *Urbanization* has become a trend, transforming big cities into megacities, and megacities into smart cities. Long story short – *smart cities* are a combination of smart governments and smart citizens, and the government–citizen bond is lifted by the presence of smart technologies. Literally, big cities are being revamped to resemble smart villages, that is, to merge humans, nature and technology. For the construction of smart cities, we asserted that there would be a need for more *adaptability* and *flexibility*. Through the implementation of a *new business model*, a fair interplay among construction actors could be ensured. And this fair *interplay* is to be sustained – bringing Copenhagen's case to mind – by a "*strategi*", an entity that could expertly orchestrate complex smart city projects: the *general contractor*.

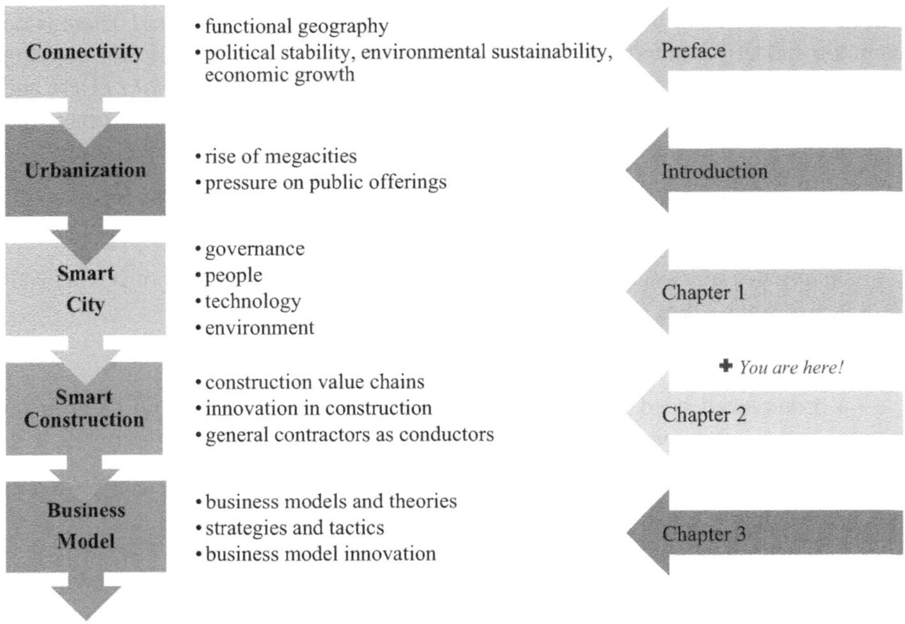

2

A General Contractor: A Maven, A Connector

This chapter assesses the role of the general contractor in construction projects; it shows how their influence on a project's final output is likely to vary based on contrasting intervention scenarios. This is particularly apt as general contractors are often disposed to play a series of roles in construction projects, rather than just one role[1].

For this reason, instead of dictating the role of the general contractor, we attempt to describe it exactly as it is, factually and objectively. *Ceteris paribus*, we demonstrate how a general contractor can make a fundamental contribution to the success of large construction projects in general and, more specifically, to smart developments.

This chapter is structured as follows. First, a description of the construction industry value chain is given, followed by a list of change agents that have enabled improved value creation in construction in recent years. A comparison between old and new construction value chains and processes is also presented. Additionally, in relation to complex construction processes (smart constructions), a concise overview of innovation in construction is provided. The role (or roles) of the general contractor is then introduced and explained. Analogies between general contractors and other construction actors are also suggested when considering the administration of smart developments. The chapter concludes by demonstrating how general contractors and music conductors are, rather surprisingly, very much alike.

---

1 Two sides of the same coin? The differences between general contractors and construction managers. Available at: https://jobsite.procore.com/construction-manager-vs-general-contractor-roles-differences/ [Accessed March 15, 2021].

2.1. New value chains in construction

A value chain is essentially a model of the corporate value-forming process and was introduced by Michael Porter back in 1985 (Porter 1985). Used for decades to understand and analyze industries, it has been revered for its ability to portray the chained linkage of activities that exists within traditional industries. By definition, value chain analysis is a process whereby companies pinpoint the primary and secondary activities that add value to their final products and services. They subsequently scrutinize these activities to reduce costs or improve differentiation (Peppard and Rylander 2006). Thus, a value chain represents the activities that companies engage in when transforming inputs into outputs.

As products and services have become ever more dematerialized and value, more than ever before, is being created within networks and alliances (rather than by solitary stakeholders), the focus has progressively shifted towards *value networks*, denoting the co-creation of value by a group of actors within the network. Referring to Santoni and Taglioni (2015), a value network is a business analysis perspective that describes social and technical resources within and between businesses; it exhibits interdependence and accounts for the overall worth of products and services.

Far-removed from the *value network approach*, a superior concept has newly emerged – the *shared value concept* – positing that value creation should not solely benefit companies and/or customers but the society in its entirety. As expounded by Porter and Kramer (2011), by (naïvely!) concentrating on improving short-term financial gains, companies have a tendency to overlook the most pressing unmet needs in the market, as well as the wider influences affecting their long-term success. As a result, they remain stuck with a nonoperational, limited approach to value creation. Furthermore, they repeatedly disregard the depletion of natural resources, the viability of suppliers and the economic agony of the communities in which they operate; all key factors in the long-term success of their business. Similar conclusions can be found in Stabell and Fjeldstad (1998). Indeed, the authors assert that a threefold value configuration analysis (*value chain*, *value shop* and *value network*) is required in order to be able to examine and comprehend the logic behind firm-level value creation, not only in the realm of the construction industry, but across a range of other industries and sectors (see Table 2.1 for an overview of alternative value configurations).

As one of the world's largest consumers of raw materials (contributing to roughly 40% of global GHG emissions)[2], the construction industry is predicted to

2 2018 Global status report – Towards a zero-emission, efficient and resilient buildings and construction sector. Available at: https://www.worldgbc.org/news-media/2018-global-status-report-towards-zero-emission-efficient-and-resilient-buildings-and [Accessed March 15, 2021].

grow steadily between 2018 and 2023 at an annual rate of 4% – in terms of market value – with major growth prospects in residential, commercial and infrastructure projects (Gawer and Cusumano 2014). This anticipated growth, along with the decarbonization imperative, have created the impetus for *sustainable construction*. Today, construction companies are becoming more accountable for their contribution to global emissions and are therefore facing societal pressure to lessen negative externalities and find practical ways to decrease their carbon footprint. Similarly, Lehdonvirta *et al.* (2009) assert that the construction sector has actually transformed into a service industry and construction companies are now seen simply as service providers. This implies that the era of technical specification and cost minimization is rapidly coming to an end. We nevertheless remain hopeful by acknowledging that there are ample technology-driven prospects for the improvement of customer value creation in construction.

	Chain	***Shop***	***Network***
Value creation logic	Transformation of inputs into outputs	Resolving customer issues	Linking customers
Primary technology	Long-linked	Intensive	Mediating
Primary activity categories	▪ Inbound logistics ▪ Operations ▪ Outbound logistics ▪ Marketing ▪ Service	▪ Problem-finding and acquisition ▪ Problem-solving ▪ Choice ▪ Execution ▪ Control, evaluation	▪ Network promotion and contract management ▪ Service provisioning ▪ Infrastructure operation
Main interactivity relationship logic	Sequential	Cyclical, spiraling	Simultaneous, parallel
Primary activity interdependence	▪ Pooled ▪ Sequential	▪ Pooled ▪ Sequential ▪ Reciprocal	▪ Pooled ▪ Reciprocal
Key cost drivers	▪ Scale ▪ Capacity utilization		▪ Scale ▪ Capacity utilization
Key value drivers		Reputation	▪ Scale ▪ Capacity utilization
Business value system structure	Interlinked chains	Referred shops	Layered and interconnected networks

Table 2.1. *Alternative value configurations (source: adapted from Stabell and Fjeldstad (1998))*

Figures 2.1 and 2.2 shown below respectively map the actual supply and value chains in construction.

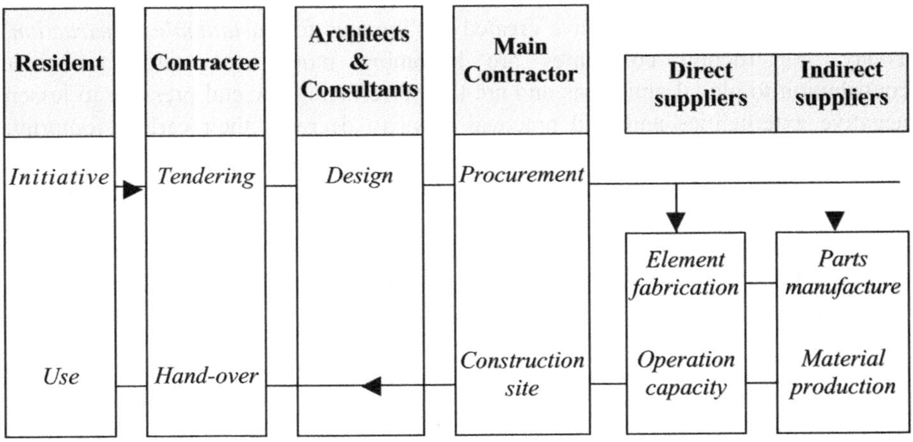

Figure 2.1. *The actual construction supply chain (source: adapted from Vrijhoef and Koskela (2000))*

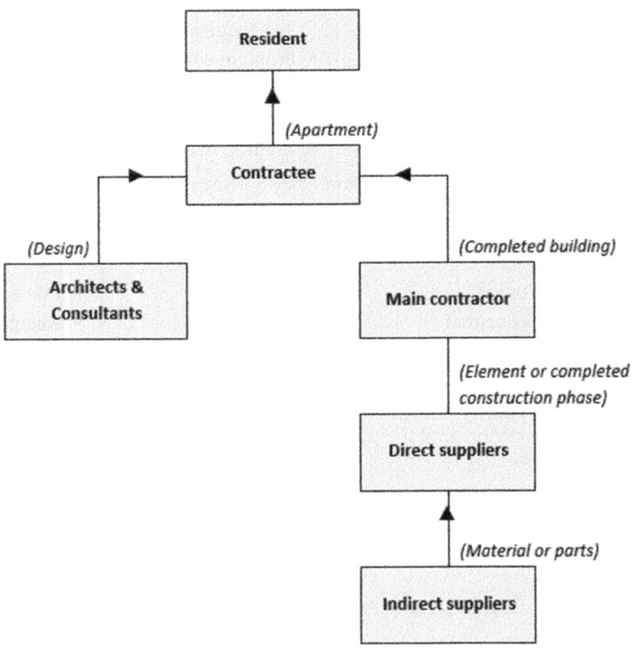

Figure 2.2. *The actual construction value chain (source: adapted from Vrijhoef and Koskela (2000))*

In the following, we briefly describe some of the change agents that have latterly given value creation in the construction industry a significant boost (see Table 2.2) (see section 1.8 for further insights into *innovation in construction*).

Additive Manufacturing Technologies (AMTs) —or— 3D Printing Methods	▪ Campbell *et al.* (2012) summarized the benefits that could ensue from the adoption of AMTs by companies, namely: customization, improved functionality, reduction of total amount of parts and esthetics; ▪ AMTs have been predicted to revolutionize the construction industry (3DRS 2015); ▪ Lim *et al.* (2012) estimated the potential advantages of AMTs for construction and came up with the following: increased freedom of design, reduction in mold costs and integrated functionality of individual components; ▪ AMTs enable cost efficient manufacturing of large, geometrically complex, unique components, using materials that are applicable to construction; ▪ AMTs are highly relevant for the construction industry as it is directly linked to logistics, customization, virtual models and manufacturing.
Virtual Reality Technologies (VRTs) —and— Augmented Reality Technologies (ARTs)	▪ Several projects have employed VRTs/ARTs in visualizing BIM[3] models (see Figure 2.3); ▪ VRTs/ARTs are changing manufacturing and enabling mass customization at an unprecedented level; ▪ Used to visualize 3D models in future construction sites, they help guide real-world construction activities (Behzadan and Kamat 2005); ▪ Consumer products enabling ARTs/VRTs have been announced by prominent ICT companies such as Samsung, Sony and Google.
Multi-Sided Platforms (MSPs)	▪ MSPs connect two or more sides of the market via online services, benefiting from network effects; ▪ They act as a foundation upon which external innovators can develop their own complementary products, technologies or services (Gawer and Cusumano 2014); ▪ Well-known examples of MSPs that disrupt traditional business models are: Uber (taxi services), Airbnb (accommodation services) and Amazon (retail services); ▪ MSPs thrive on data-driven customer understanding, enabled by faster innovation capabilities and greater profits than the general industry average (Weill and Woerner 2015).

Table 2.2. *Change agents in the construction industry (source: created by the author from various sources*

Without doubt, new knowledge and high-tech have been (and still are) creating competence-based growth and breakthroughs in some of the nascent fields of society today: artificial intelligence, robotics, navigation and data analytics. We believe the blanket adoption of these applications would eventually have the ability to transform the entire value chain of construction (see Figures 2.3 and 2.4).

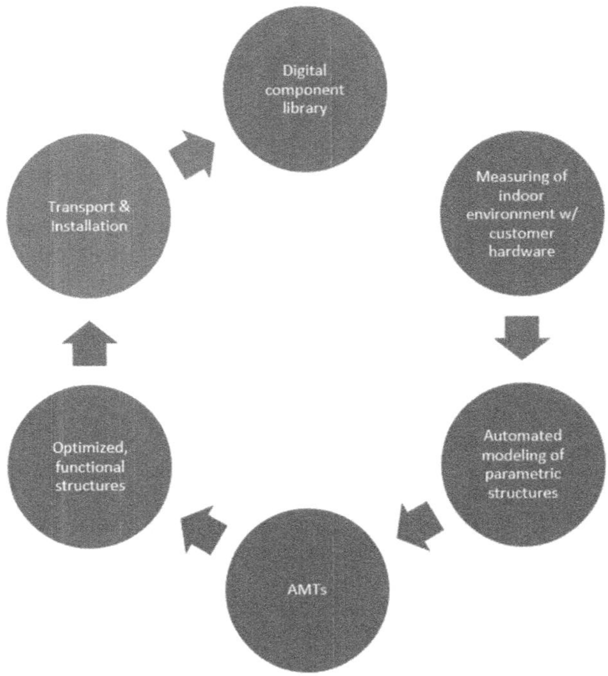

Figure 2.3. *The future of construction process (source: adapted from Virtanen* et al. *(2014)). For a color version of this figure, see www.iste.co.uk/karam/general.zip*

When looking at the construction value chain (actual vs. future), it is easy to see that the variances occur at the beginning and during the later stages of the chain.

Economically, this suggests an incremental increase in value in very small steps to a hasty upsurge in value in one jump. If we consider Figure 2.4 (relative to Figure 2.2), the *central operator*[3] has noticeably more control over the manufacturing process in terms of

3 Also known as the *online contractor* – the equivalent of a *general contractor* – the term used in this book. Sections 2.4 and 2.5 provide more insights into the roles played by *general contractors*. A detailed explanation of the *general contractor* business model for the construction and management of smart cities is available in Chapter 4.

both scheduling and price determination. Moreover, the logistics and materials held in storage are reduced, limiting the amount of resources tied to the process. For clients (residents) and designers, the ability to customize is amplified; thus, fewer conciliations are needed to fit the design to the offerings of manufacturers. In the final parts of the chain, the central operator turns into a central stakeholder. For small-sized operations, the roles of designer, manufacturer and contractor may be merged into one. This is made possible by the clients who provide digital data of the site and make use of IoT; this enables clients to make some design decisions and to make use of AMTs for the production of more complex components. Overall, the number of stakeholders involved in the construction process is reduced, resulting in fewer communication and procurement steps.

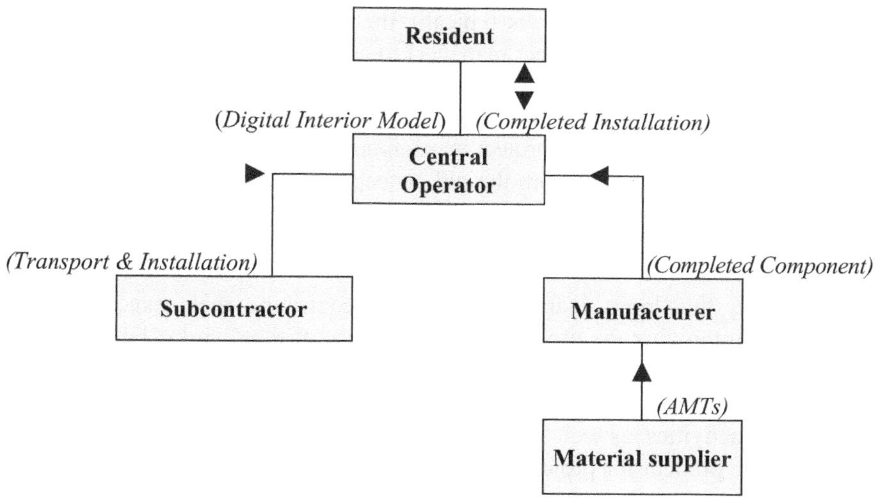

Figure 2.4. *The future construction value chain (source: adapted from Virtanen* et al. *(2014))*

Currently, a number of concerns surround the analysis of future construction value chains. In particular, these center on the fact that the predictions made – which are often deemed unsatisfactory – account for technical developments only. In addition, it seems very optimistic to assume that a single value chain could serve all situations encountered in the construction industry. For example, retrofitting a single apartment and building a new residential area are inherently distinct projects, of completely different scales; the producer–customer relationships involved also differ significantly. Thus, it is inconceivable that a common value chain exists, able to precisely fit to all types of construction projects (Virtanen *et al.* 2014).

2.2. The construction project lifecycle

Every stakeholder involved in the process of scheduling, planning, sponsoring, building and operating physical facilities related to construction projects gains different viewpoints over time on project management for construction (Figure 2.5). Indeed, contributing expert knowledge could be very advantageous, particularly when it comes to large and complicated projects (smart developments). In the same way, it is of foremost importance for all stakeholders to understand how the constituent parts of a construction process are connected together.

As conveyed by Hansen and Birkinshaw (2007), poor coordination and communication between stakeholders could eventually result in waste, excessive costs and unwelcome delays. It is typically the responsibility of the project owner (and to a lesser extent the project manager) to assure that such issues do not occur. Egbu (2008) found that the implementation of the project owner's viewpoint would help stakeholders to focus on the completion of the project by paying close attention to the details of the process of project management for construction projects. Indeed, this would be a move away from the old concept of making decisions based on the bygone roles of stakeholders involved in the project (e.g. project owners and managers, architects, general contractors and others).

Following this logic, stakeholders would contribute their expertise through opinions on improving the productivity and quality of their works. For construction actors to be able to make substantial improvements, they must first know the construction industry, its working environment and the institutional constraints affecting its activities, as well as the nature of project management (Carassus 2004). It is therefore prudent for project owners to have a solid understanding of the whole construction process in order to maintain control of the quality, suitability and budget of the finalized project.

The typical CPLC – represented in Figure 2.6 – consists of five major phases, explained in Table 2.3. In addition, a complete view of a construction management process is shown in Figure 2.7.

In practice – depending on the nature, size and urgency of the project – these development phases may not always be sequential as some may require iteration and others may be carried out across overlapping timeframes. Further to this, it is worth asserting that project owners are likely to establish in-house capacities to handle the works in every phase of the construction process. Yet, if impossible to do so, they tend to outsource part – or all – of the jobs by seeking external expertise. This expertise may come from a general contractor or a construction manager, for instance, who could conduct and monitor the works carried out at every stage of the project.

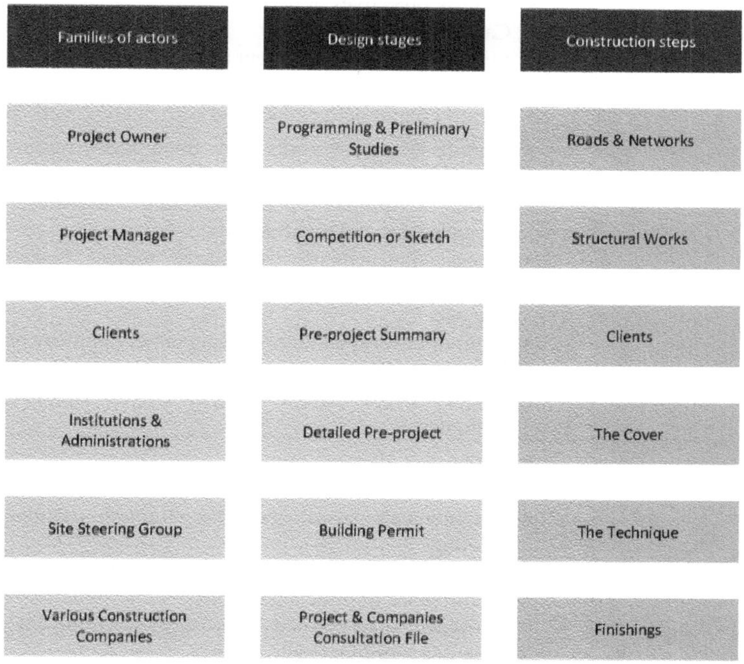

Figure 2.5. *The building blocks of a construction project process (source: adapted from Direction des Immobilisations and Cloutier (2005). For a color version of this figure, see www.iste.co.uk/karam/general.zip*

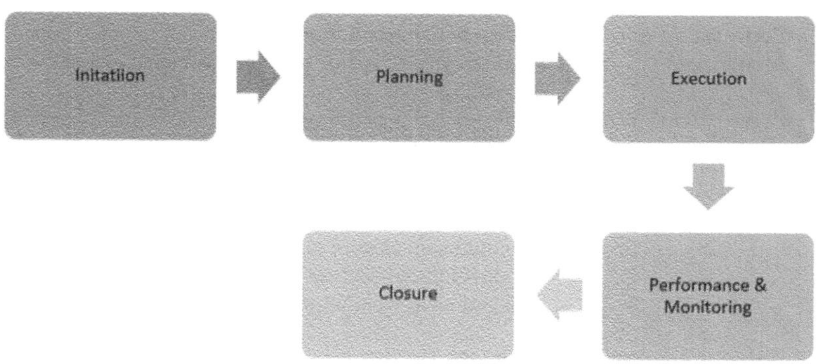

Figure 2.6. *Major CPLC phases (a) (source: compiled by the author[4]). For a color version of this figure, see www.iste.co.uk/karam/general.zip*

4 Top five project management phases. Available at: https://project-management.com/top-5-project-management-phases/ [Accessed March 15, 2021].

I	Initiation	▪ Identify objectives or needs ▪ Carry out feasibility study ▪ Initiate project to implement approved solution/s ▪ Appoint 'project manager' ▪ Identify major deliverables and participating work groups
II	Planning	▪ Identify all of the works to be done: tasks, resources and strategy ▪ Build a project plan outlining activities, tasks, dependencies and timeframes ▪ Scope Management: Coordinate project budget (cost estimates for labor, equipment and materials) by project manager ▪ Monitor cost expenditures during project implementation ▪ Provide quality targets, assurance and control measures ▪ Define acceptance plan, listing the criteria requested by customers
III	Execution	▪ Put project plan into motion ▪ Perform assigned tasks (by construction actors) ▪ Communicate work progress through regular team meetings ▪ Compare progress reports with project plan to measure the performance of project activities ▪ If needed, plan corrective measures (to bring the project back to the original plan) ▪ Keep key stakeholders (including project owners) informed about the project's status ▪ Indicate possible end point in terms of cost, schedule and quality of deliverables in status report ▪ Review deliverables for quality, against acceptance criteria
IV	Performance & Monitoring	▪ Measure progress and performance of project activities (in tandem with execution phase) ▪ Track project activities with project management scheduling
V	Closure	▪ Provide final deliverables to 'client' ▪ Project documentation handover to the business ▪ Terminate supplier contracts ▪ Release project resources ▪ Communicate closure of the project to stakeholders ▪ Conduct lessons-learned studies to examine what went well and what went wrong

Table 2.3. *Major CPLC phases (b) (source: compiled by the author[5])*

5 Ibid.

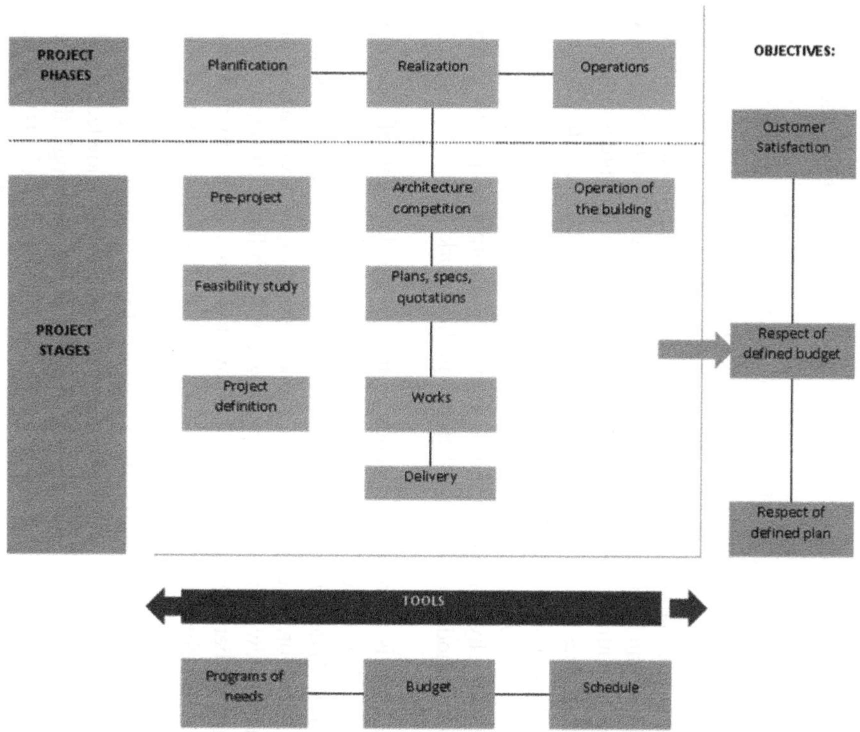

Figure 2.7. *The construction project management process (source: created by the author). For a color version of this figure, see www.iste.co.uk/karam/general.zip*

2.3. Innovation in construction

Innovation could be a key source of competitive advantage for construction companies (Slaughter 2000). Yet the construction industry is an example of a sector in which conventional measures do not convey the true extent of the innovative activity that is actually occurring – and is often seen as one of the less inventive sectors (Barrett *et al.* 2007). We believe this observation is somewhat unwarranted because much of the inventiveness remains out of sight as it is co-developed at the project level. As a project-based industry that is largely disjointed, types of innovation in construction differ from those of others. Though a significant amount of research has been conducted to date, more is needed to investigate different types of original activities carried out throughout the CPLC.

As per Gann and Salter (2000), project-based companies need to manage their project processes and business models effectively given their resources are embedded within both project- and company-levels. "It is the integration of these two sets of resources that enables the firm to be competitive" they added. On another note, Brusoni *et al.* (1998) indicated that business models are ongoing and repetitive, whereas project processes are more likely to be temporary and unique. This is the reason why companies must continually incorporate project experience into their business models to ensure coherence in their work.

Currently, there is a growing tendency to take a broader view of innovation in construction, one that reflects the many ways in which innovation occurs in practice. Damanpour's (1992) definition of innovation is suitably inclusive: "it is the adoption of an idea or behavior, whether a system, policy, program, device, process, product or service, that is new to the adopting organization". Phillips (1997), on the other hand, distinguished between technological innovation and organizational innovation. For the author, the former incorporates substantial technological improvements in products and processes, whereas the latter involves major changes in organizational structures. Furthermore, Slaughter (1998) suggested five innovation models ranging from incremental innovation – slight changes based on current knowledge and skills – to drastic innovation, for example a technological breakthrough that alters the entire nature of an industry. Subsequently, Blayse and Manley (2004) emphasized that construction is partly manufacturing and partly service industry and, crucially, that the characteristics of innovation in the service industry are different from those in the manufacturing industry.

As a weighty economic variable, the measurement of innovation has attracted a great deal of attention among academic circles in recent years. However, due to the intricacies of construction processes in general, evaluating innovation remains a difficult task to accomplish. In general, construction is a very diverse sector and there is no precise way in which innovation occurs. As per Lansley (1996), the occurrence of innovation within the construction industry is often characterized by the continuous adoption of new practices because of advances in technological processes and business models. For Barrett *et al.* (2007), innovation can be observed at the sector, business or project level; innovation is most visible at sector level and least visible at project level. The construction sector is an interlocked system involving many stakeholders, among which are clients who have the potential to act as innovation creators. In this context, clients could spur innovation in construction by:

– exerting pressure on supply chain partners to improve overall performance (Gann and Salter 2000);

– helping them to devise strategies to cope with unforeseen changes[6];

– requesting high standards of work (Barlow 2000);

– stipulating explicit and forward-looking requirements for a project (Seaden and Manseau 2001).

Knowledge provision, operative leadership and dissemination of innovations are also key parts that clients could play in this respect (Egbu 2008).

As previously noted, existing literature on innovation in construction neglects project-level innovations, largely because of difficulties encountered when trying to monitor the various activities carried out by stakeholders in each phase of the project. As Marceau *et al.* (1999) observe, the management of innovation in construction is complicated by the sporadic nature of project-based production which regularly includes broken learning and feedback loops. For this reason, we believe that a deeper understanding of the different types of innovative activity that occur during a CPLC is necessary to enable their effective application and management, consequently allowing for value creation that could ultimately benefit anybody and everybody.

2.4. The general contractor

Following copious amounts of reading, and after having carried out several investigations in my assiduous search for some real and relevant information about general contractors, I can now resolutely corroborate that the existing academic

6 Ibid.

literature on the matter is quite limited. This is exemplified by the fact that all of my attempts to gain insight into the role of general contractors in general – and in construction management services, specifically – have unfortunately come to a dead end.

However, in order to develop the subject further, I have conducted some further analyses and compiled ample real-world information about general contractors. The focus is on what they do, how they compare to other construction actors and – applying Freakonomics-style thinking[7] – what they have in common with music conductors, and why this is important. The following is based on secondary data, taken from specialist magazines, websites and blogs, alongside talks held with peers who – like myself – are experts in the building and construction industry. At various points, I intervene by voicing my own standpoint to explain some technical nuances that would otherwise have remained elusive.

2.4.1. *Roles and responsibilities*

GCs (General Contractors)[8] could assume many different roles and positions when working on a construction project. Naturally, the responsibilities of GCs vary depending on the size and complexity of the project[9]. Indeed, GCs are responsible for a number of details over the course of a given construction project. Finding the right talent to get the job done is clearly one of their most important tasks, though not the only one. Today, there is much discussion about the varied roles that GCs could play in a project (Figure 2.8). Indeed, there are so many instances of GCs who spread themselves too thinly in an attempt to satisfy the various needs of projects and clients (see Table 2.4). Interestingly enough, a crucial move in that direction is currently taking place[10].

By definition, a GC is a stakeholder – an entity or a person – who manages the construction work in its entirety, supervises daily activities at construction sites and ensures all construction works are carried out properly. Moreover, a GC supplies the labor, building materials, machinery, raw materials and all other kit necessary for the successful completion of the work in general. For industry experts, the GC is a

7 For more information, see: http://freakonomics.com/archive/ [Accessed March 15, 2021].

8 Sometimes referred to as: *Main Contractors*.

9 Important facts you need to know about a general contractor. Available at: https://construction.laws.com/general-contractor [Accessed March 10, 2021].

10 Who is a contractor (definition)? Available at: https://www.letsbuild.com/blog/contractor-role-duties [Accessed March 10, 2021].

"central operator" who provides tailored solutions to clients, and expertly manages worksites, tradespeople, deadlines and budgets[11]. Typically, GCs are hired via a bidding process and often the lowest bid for the least qualified entity is selected for the job. The GC's set price is based on the contract terms and the construction drawings. Should the GC spend less than the bid, then they profit on the differential. On the contrary, if the GC bid is over budget, then the project owner would then be over budget as well (at the outset) and would need to adjust it accordingly. This could involve providing extra money to cover the costs, changing project specifications, or reducing the scope; it then becomes apparent why the relationship between the GC and the owner is often regarded as competitive (Yang *et al.* 2010).

Furthermore, GCs do not always have the required expertise to complete all of the construction work themselves. For this reason, they maintain a network of specialist subcontractors (they have their own construction value chains!). GCs are the *big picture thinkers*[12]; they are involved in construction projects from the inception phase to the delivery of the final product, whereas subcontractors, for instance, step in only to fulfill particular tasks: concrete formulation, plumbing, electricity, carpentry, etc.

One of the many roles assumed by a GC is to maintain direct contact with clients to keep them informed about progress being made, as well as with other construction actors (architects, engineers and others) throughout the whole CPLC.

In addition, GCs often handle hiring and are responsible for instructing those new to the job. In any project, there are a number of codes, laws and regulations that GCs must abide by at all times. They must follow applicable labor laws, including organized union contracts under which their staff work and must be aware of safety regulations for various tasks and equipment operators. The GC is therefore responsible for guaranteeing that all of these situations are handled properly, whether that be securing medical attention for an incapacitated worker, locating backup equipment or expediting a supply order. When progress is delayed, the GC must look for ways to get the project back on track. This is imperative because delays in construction projects are typically very costly (Ramanathan *et al.* 2012).

11 What is the role of general contractors in construction work? Available at: http://www.accconstruction.ca/what-is-the-role-of-general-contractor-in-construction-work/ [Accessed March 10, 2021].

12 Interview with Raymond Vigneau, owner of Metal Building Contractors, Inc. Available at: https://home.howstuffworks.com/home-improvement/construction/planning/why-hire-a-contractor1.htm [Accessed March 10, 2021].

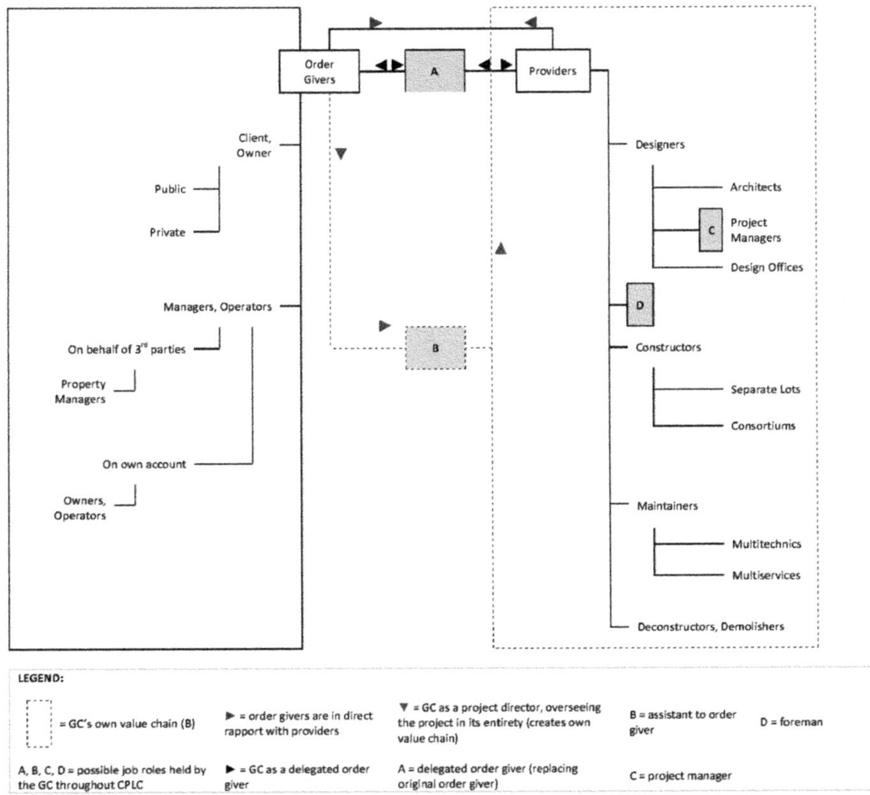

Figure 2.8. *Construction sector organigram (source: created by the author. For a color version of this figure, see www.iste.co.uk/karam/general.zip*

SCENARIO:	A	B	C	D
Job role:	Delegated Project Owner	Assistant to Project Owner	Project Manager	Foreman
Involvement:	More	←	→	Less
	Early stages of value chain	←	→	Later stages of value chain
Intervention:	Such roles are now being endorsed by France-based construction operators (specifically sizeable general enterprises) due to the copious operational and managerial issues they are facing while executing full-scale, smart developments. This is indeed the case for *Bouygues Construction* and *Vinci Construction*.		As conveyed earlier, GCs play several roles and hold different positions along the construction value chain. Such roles could be performed by GCs at any time and in any project, regardless of its type or size.	
Added value[2]:	Very High	High	Mid-to-High	Low
Level of challenges encountered:	Nil	Insignificant	Moderate	Extraordinary
Selective list of challenges encountered:	None	Trivial operational issues	One or more of the challenges listed below could be encountered under scenarios C & D: ▪ Disregard of defined plans ▪ Cost or budget overrun ▪ Scheduled timeline overrun ▪ High risk of technological obsolescence upon delivery of project ▪ Liability concerns as to who is responsible for what ▪ Criticism of final product (by clients) ▪ Customer dissatisfaction	

Table 2.4. *General contractors' roles (source: created by the author. For a color version of this table, see www.iste.co.uk/karam/general.zip*

In terms of the contract to be signed, project owners may possibly add a "no-damage-for-delay" clause, thus shifting the burden of potential delays to the GC. We can therefore summarize by attesting that:

– it is very complicated to establish a precise definition for the role/s and duties of a GC because of the range of services that they are capable of providing;

– however, it is quite easy to ratify that GCs are both mavens (experts in the field of building and construction) and connectors (orchestrating the work of all construction actors).

Lastly, it is important to point out that the advent of high-tech (IoT, big data and others), and the increase in smart constructions and initiatives around the world – coupled with the issues the construction sector has been facing in terms of project delivery – have together paved the way for defining (and endorsing) a new job description for GCs.

2.4.2. *General contractors versus other industry actors*

GCs, in contrast to other construction actors, have the potential to play several roles and hold various positions – A, B, C or D – within the construction value chain. In view of that, their level of contribution to the overall success of construction projects is likely to vary based on whether they are involved at the very early stages, or further down the chain. The challenges faced by GCs are, by the same token, likely to vary based on intervention stage/time. Put simply, the more involved GCs are (i.e. the earlier they become involved) in large-scale construction projects, in particular, the less likely they are to face challenges at later stages. Furthermore, the less likely it is that timings and costs (among other aspects) would alter as the construction work progressed.

As shown in Figure 2.8 – and then expounded in Table 2.4 – a GC's added value would be hypothetically maximized under scenarios A and B. This is primarily because they will be given the opportunity to create their own value chain and to manage the different parts of the project on their own, in order to meet requirements in the best way possible. Within this framework, GCs would have complete control over all aspects of the project, would manage all construction services as they see fit and consequently be able to deliver the project on time and within budget. If the opposite should happen, however, i.e. should any issues arise (resulting in deviation from the original plan) then they would be the only ones to assume the entirety of ensuing costs. Under scenarios C and D, however, the contribution made by GCs in terms of the success (or failure) of a construction project is only marginal. They therefore cannot be held responsible for any operational snags they might encounter, except for those directly within their mandate.

We continue our investigation by considering the roles of two industry actors, namely: *construction managers* and *project managers*, in terms of what they do and how they compare to GCs.

2.4.2.1. *Construction managers*

A large construction project produces a veritable cacophony as a result of the substantial number of people involved in it. And for those involved to be able to create beautiful tunes together, a conductor is needed for guidance and direction. For owners, attaining the desired outcomes on a full-scale project is a huge challenge that they have to address on a daily basis. Coordinating the many subcontractors, suppliers and other personnel – among numerous aspects – is a balancing act that requires attention and fortitude. We believe that the people who can accomplish these key tasks fall into one of two roles: GCs (General Contractors) or CMs (Construction Managers).

The ultimate goal of both roles is essentially the same, that is, to put all of the pieces of the puzzle together and execute the design according to what the project owner wants and what the regulations demand. Indeed, GCs and CMs have the same objective of completing the project to the satisfaction of the owner[13]. However, have you ever considered the disparities that exist between GCs and CMs? It may seem as though they have the same job but how similar are their jobs really? Despite the apparent similarities between the two, a GC and a CM go about their roles slightly differently as each has their own set of unique financial structures and duties; thus they each have their own strategies for safe and efficient execution of construction projects in general. Other key differences relate to their organizational structures, how they are selected for projects along with their entry point, as well as their associations with project owners[14].

Commonly, GCs and CMs are considered to be the main contractors on the job, bringing with them different apporaches to construction projects. The decision to appoint either one of them – *chief conductor of the orchestra* – depends upon owner preferences, which could vary on a project-by-project basis. Indeed, for everyone to sing from the same song sheet, in harmony with the owner, seems to be rather fundamental (see section 1.10 for a rogue analogy between GCs and music conductors). But why do project owners tend to favor CMs over GCs? The answer is simple. Owners prefer to work with CMs because of the personal relationships they have developed with them over the years which has resulted in an appreciation of, and familiarity with their work.

---

13 Two sides of the same coin? The differences between general contractors and construction managers. Available at: https://jobsite.procore.com/construction-manager-vs-general-contractor-roles-differences/ [Accessed March 12, 2021].

14 Construction managers vs. general contractors: What is the difference? Available at: https://esub.com/construction-manager-vs-general-contractor/ [Accessed March 12, 2021].

Nonetheless, it is not unusual for a GC to act as a CM. In fact, a GC who has handled a few projects for a given project owner and has subsequently established a relationship with that owner, could be asked to work on a new project as a CM. The drive for doing so is less about money and more about trust and a preference for the style and substance of that the GC. This overlap between the two roles, i.e. GCs acting as both a GC and a CM (we believe!) has a best-of-both-worlds benefit. This is particularly true for the owners in that it reduces the possibility of any potentially confrontational exchanges occurring between them and GCs, while upholding the close ties between GCs and their subcontractors[15].

Table 2.5 shows the CM–GC (Construction Manager–General Contractor) project delivery method as an alternative to other methods: D–B–B (Design–Bid–Build) and D–B (Design–Build).

Level of Design Completion	*Fully developed scope*		***Jointly developed scope***			*Undeveloped Scope*
Project Delivery Method	*D-B-B*		***CM-GC***			*D-B*
Procurement Procedure	*Low Bid*	***Best value***		***Qualifications-based***		*Sole Source*
Payment Provision	*Unit price, lump sum*		***Max. Price***		*Cost Plus Fee*	*Cost Reimbursable*

Table 2.5. *Project delivery system spectrum (source: adapted from Gransberg and Shane (2015))*

According to Antoine (2017), use of CM–GC and D–B methods has increased significantly in recent years. These methods are hands-on routes for curbing construction project durations, delivering projects more expertly and speedily, and establishing early cost certainty during project delivery. Equally importantly, Le Masson and Weil (2009), and Rosati and Conti (2016) explain that the CM–GC method – being either best value or qualifications-based – allows owners, designers and/or architects, as well as GCs, to work together to tweak the scope of construction projects and make rational project management decisions.

---

15 What is the difference? General contractors vs. construction managers. Available at: https://www.suretybondsdirect.com/educate/general-contractor-vs-construction-manager [Accessed March 12, 2021].

In addition, Flyvbjerg (2014) acknowledged that, under the CM–GC method, GCs are not nominated – at least not anymore – through passive bidding processes but based on valid credentials, past performances and success records.

2.4.2.2. *Project managers*

In construction, project completion is always the ultimate goal; however, achieving this in an efficient and timely manner is sometimes challenging. For this reason, a PM (Project Manager) is often required and hence, we ask: What do PMs actually do? And how do they compare to GCs?[16] Both GCs and PMs play a key role in construction projects. While PMs operate at a slightly higher level, coordinating the different entities involved in the project, GCs also play a coordinating role among different employees, specialists and subcontractors they work with to ensure project success. Thus, while both actors carry an obligation to bring together the different bodies that operate under them, GCs and PMs function at different levels in the construction management process. This can be a source of confusion for people trying to figure out what sets these two management positions apart.

For small- to medium-sized construction projects, the owner (client) could possibly choose to work with a GC who will not only perform much of the construction work but will also help coordinate between different parties involved in the process and serve as the main point of contact for the client. Nevertheless, according to the status quo, for larger projects the coordination of different entities involved in the project would be the duty of the PM.

Despite some similarities in their roles (as you may have guessed by now!), PMs and GCs remain poles apart. Generally, PMs are accountable for high-level coordination of construction projects. Unlike GCs, who are chiefly responsible for the actual physical construction of a project (partially or wholly), PMs work closely with all parties involved in a project to ensure that the owner's goals have been met, the project has stayed within budget and the project is delivered on time. In their role PMs serve as team leaders and primary liaisons with the client. As for clients, they tend to work closely with a project management team so that the scope and goals of a project are clearly defined; in other words, developing a strategic plan. When creating a strategic plan, the PM and the client join forces to define the scope of the project and the desired timeline and budget. Aside from drafting a budget that aligns with the client's expectations, PMs are also responsible for implementing that budget and ensuring the project remains on (or below) budget throughout the whole CPLC. This is key as because large construction projects are well known for going over-budget.

---

16 General contractor vs. project manager. Available at: https://gillilandcm.com/2018/12/06/general-contractor-vs-project-manager/ [Accessed March 12, 2021].

GCs	• Typical business entities; • have their own complement of staff in the form of a pool of subcontractors; • specialize in certain types of construction and in certain construction sectors; • chosen via bidding processes and involved in construction and in the daily direction and operation of projects; • ensure that all works are completed properly and on time; • complete physical works onsite with the help of their teams of construction workers; • could act as CMs and be involved in construction projects early on as an advisor (this is factual given there is an established rapport between them and the owners; in this case, GCs no longer need to submit competitive blind-bid proposals but realistic ones based on insight into the development of the design).
CMs	• Could be an individual, group of people or an organization; • selected based on qualifications and experience (rather than bids); • pay is fee-based thus they receive a percentage of the total project cost (no competition for profits like with GCs); • involved at the very beginning of the project providing input on the design and working with subcontractors to provide accurate costs and timeframes; • maintain collaborative, win-win rapport with the owners and tend to work exclusively for them; • on staff with CMs are estimators, accountants or other professionals with duties that come into play before, during and after a project; • are involved during pre-construction and work with the design architects; • work with onsite managers who handle the projects during construction; • advise owners and lead construction workers; • responsible for setting/keeping to schedules and monitoring finances.
PMs	• Involved in all aspects of the construction project, including pre-construction activities, construction administration and post-construction; • understand client objectives and priorities and ensure that all project consultants are in line with these objectives; • manage staff according to target capacity, budget, timeframe and quality of project; • on-site throughout the entire project and, unlike GCs and CMs, they oversee it from pre-construction to closure; • in some cases they supervise CMs and/or GCs on behalf of clients.

Table 2.6. *General contractors vs. construction managers vs. project managers*[17]

17 The difference between a project manager, construction manager and general contractor. Available at: https://watchdogpm.com/blog/the-difference-between-a-project-manager-construction-manager-and-general-contractor/ [Accessed March 12, 2021].

Ceteris paribus, a competent PM would ultimately add value to a given construction project by defining a comprehensive budget that incorporates all expenditures, capital and operational, and by sticking to that budget over time. Furthermore, PMs oversee the execution of projects and assemble the teams needed to complete them. Assembling teams is often carried out by issuing a *request for proposal* to the entities concerned: architectural, engineering and general contracting companies. During this process, PMs are expected to perform a certain level of due diligence to ensure the companies chosen have experience with projects of a similar type, as well as the necessary qualifications and KB (Knowledge Base)[18] to successfully complete the assigned project. Subsequently, PMs oversee contract negotiations to guarantee that the scope of the project, budget and timeline are clearly understood by all parties in the project team (see Table 2.6).

2.5. How do general contractors resemble music conductors?

Now we turn to music! Conducting is the art of guiding the concurrent performance of several players[19] through hand gestures, usually with the aid of a baton[20]. In other words, conducting is a means of communicating directions to musicians during a performance[21]. Actually, there are many formal rules that state how to conduct properly, as well as a range of conducting styles that tend to vary according to the education and complexity of the conductor. In general, a conductor's main aim is to unify performers, set the tempo, ensure well-timed entries by ensemble members, execute clear beats, listen analytically to shape the sound of the ensemble and control the pacing of the music.

Communication (as you might expect!) is nonverbal during live performances and yet it is verbal during rehearsals; conductors may stop the playing of a piece to orally request some style changes or ask for adjustments in the tone of a certain section.

18 A KB is a technology used to store complex structured and unstructured information used by a computer system.

19 A full-scale orchestra playing a symphony comprises over 90 musicians, while a smaller orchestra playing a chamber piece, for instance, includes a maximum of 45 musicians.

20 Louis Spohr claimed to be the first to introduce a conducting baton to England in 1820 but reports indicate that Daniel Turk conducted the Halle Orchestra with a baton in 1810, 10 years earlier.

21 The conductor of an ensemble. Available at: https://www.liveabout.com/what-is-a-conductor-2456662 [Accessed March 14, 2021].

In an orchestra, the smallest instrument is the piccolo which is a half-size flute. The largest family of instruments is the strings and these come in different sizes: the violin (the smallest), followed by the viola and the cello and then the double bass (the largest). Figure 2.9 shows how an orchestra is arranged; the first chair violinist, usually regarded as the best, sits to the music conductor's left, adjacent to the audience. They are the "concertmaster" – the leader of the violin section – who, in smaller orchestral settings, has the potential to lead the ensemble independently hence, to act as a music conductor.

NOTE.– We hope by now that you are able to detect the hidden similarities between the roles played by conductors and concertmasters (owners, PMs, CMs and GCs), and how these roles will likely change according to the size of the orchestra (construction project)!

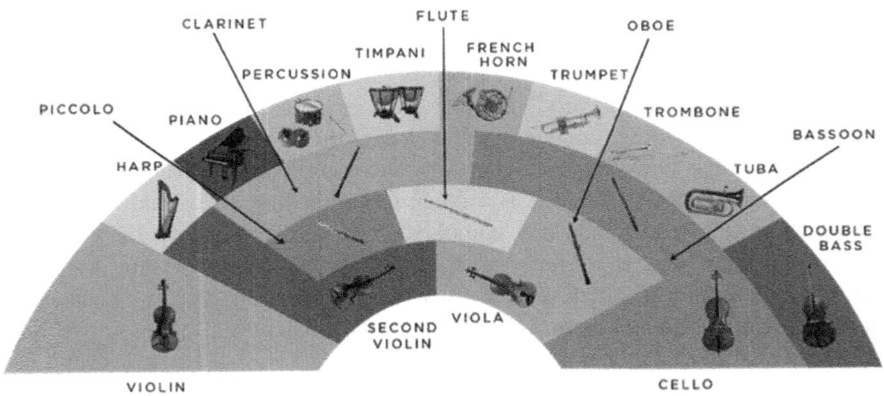

Figure 2.9. *Sections of the orchestra*[22]

In the music industry, a conductor is occasionally referred to as a director or chief conductor. Other conductors, usually those who lead choirs, are at times referred to as choral directors, chorus masters or choirmasters. Conductors of concert (military or marching) bands, on the other hand, hold a number of titles, for example, band directors or bandmasters. And senior music conductors are routinely called maestros. BOOM! GCs are given several names too: main contractors, prime contractors, head contractors and others.

22 Sections of the orchestra. Available at: https://www.thinglink.com/scene/682291240225472514 [Accessed March 14, 2021].

Now that we know just a little bit more about orchestras and music conductors, we may proceed by asking the following questions[23]: Why are orchestras structured in the way that they are? Why are the oboes and tubas positioned at the front? Why do flutes and violins not swap positions? Why do trombones and French horns not sit right up front with the music conductor? In fact, (we believe!) there is a good reason for why pretty much everything we experience (in life) is the way it really is; the seating arrangements of an orchestra are no exception. In an article[24] for The Florida Times-Union, Courtney Lewis – the Jacksonville Symphony Music Director – gave a concise explanation for why players sit where they do. According to Lewis, "orchestras owe their present-day configuration to two phenomena (surprisingly!): industrialization and urbanization [...] The current seating arrangements are a fairly contemporary development".

Until the 20th century, the violins – first and second – were seated opposite one another, for the most part to create a *stereo effect* with the two sections playing off one another. A few years later, the influential conductor – *Leopold Stokowski*[25] – appeared. Stokowski was an alchemist; he tried seating the orchestra in every conceivable way in an effort to find the perfect amalgam of resonances. This is actually the responsibility of the central operator who tries to bring all construction actors together, in order to create beautiful melodies rather than mishmashes of conflicting sounds.

But it was only in the 1920s that Stokowski made a change that stuck. Trusting that all violins should be placed together to help the performers hear one another better, he proceeded by organizing the strings from high-to-low and from left-to-right. The *Stokowski Shift* – as it then became known (a new orchestral structure)[26] – was espoused by orchestras all over America. In the UK, the same arrangement was favored a few years later, leading to its general adoption across the country. This structure, though unpopular at first, did gain influence and power over time. And the outcome? (You guessed it right!). Composers from around the world started writing pieces of music that took advantage of this arrangement. (Hooray! New business structures seem to be catching.)

Let us refer back to conducting; a conductor must understand all the elements of musical expression (tempo, dynamics and articulation) and have the ability to

23 Why is the orchestra seated that way? An explanation. Available at: https://www.wqxr.org/story/why-orchestra-seated-way-explanation/ [Accessed March 14, 2021].

24 Ibid.

25 He served as the music director of the Philadelphia Orchestra.

26 The equivalent of a new business model – or value chain – in construction.

communicate them effectively to an ensemble. This is also true for those assigned to lead construction projects! The ability to communicate nuances of phrasing and expression through gestures is also a prerequisite to becoming a competent conductor[27]. This is indispensable because dynamics in music are often signaled with hand gestures and communicated by the size of the conducting movements, with larger shapes representing louder sounds. Moving forward, the indication of entries – that is, when a performer (or section) should begin playing – is called *cueing* (corresponding to when-and-how each construction actor is supposed to intervene throughout a CPLC). A cue indicates the exact moment of the coming beat so that all players responding to the cue can start playing in chorus. A cue is vital for indicating when performers should change to a new note. And cueing is achieved by engaging the players before their entry and executing a clear preparation gesture, often directed towards the specific players[28]. (The latter statement explains why strategic project management plans are so important.)

To sum up, conductors act as leaders of the orchestras they conduct; they choose the works to be performed and study their scores[29]. Nevertheless, within particular frameworks, they could also handle a series of executive tasks (e.g. schedule practices, plan concert seasons, hear auditions and choose participants, as well as promote their ensemble in the media). Indeed, the role/s that a music conductor could potentially play vary greatly between different conducting positions and ensembles. (We hope you are able to keep count of the numerous similarities between GCs and conductors.)

In some cases, a conductor could be the musical director of the symphony, choosing the program for the entire season. In other cases, the conductor could attend some or all of the auditions for new members of the orchestra to ensure that the runners have the required playing style and tone, and meet the highest performance standards. Within different settings, conductors could be hired simply to prepare a choir for a number of weeks, which would then be directed by another conductor. A number of other conductors, on the other hand, could have a

27 What is the career path to becoming a conductor? Available at: https://www.connollymusic.com/stringovation/career-path-to-become-a-conductor [Accessed March 14, 2021].

28 What is a music cue? Available at: https://www.mediamusicnow.co.uk/information/glossary-of-music-production-terms/what-is-a-music-cue.aspx [Accessed March 14, 2021].

29 An orchestral score shows all parts of a large work, with each part on separate staves in vertical alignment, and is for the use of the conductor. Thus, the conductor could see at a glance what each performer should be playing and what the ensemble sound should be.

substantial public relations role, giving interviews to the local news channel and appearing on talk shows to promote the approaching season's concerts. It is clear that music conductors, very much like GCs, play different roles and assume different positions along the musical value chain.

Once again, we wonder: Why is there even a music conductor at all?[30] Traditionally, all orchestras played without conductors and were directed by concertmasters. It was only at the start of the 19th century that orchestras became big enough for a conductor to be necessary. Indeed, year after year, orchestras have turned into very large affairs. Similar to stakeholders involved in large and complex construction projects, every musician in a full-scale orchestra could have differing ideas in terms of cues, tempo and how best to play their instrument. Therefore, the main purpose of a music conductor is to interpret the music and allow the performers to get through the piece being played seamlessly, without stopping. Furthermore, a conductor ensures that all instruments are impeccably orchestrated so that none of them ends up being drowned out.

Before closing, let us first imagine how the preceding paragraph would read (or sound!) if we were talking about GCs rather than conductors. By swapping just a few words only, we get the following narrative:

> Traditionally, all construction projects were administered without GCs and were directed by project owners. It was only recently that construction projects became big (smart) enough for a GC to be necessary. Indeed, year after year, construction projects have turned into very large affairs. Similar to performers in contemporary orchestras, every stakeholder in a full-scale construction project could have differing ideas in terms of planning, scheduling, budgeting, and construction management services. Therefore, the main purpose of a GC is to oversee construction works and allow for stakeholders to get through the project being built seamlessly, on time and within budget. Furthermore, a GC ensures that the tasks fulfilled by construction actors are impeccably orchestrated so that none of them ends up being drowned out.

Astonishing, right? In any case, regardless of whether you appreciate this fabricated reconciliation between GCs and music conductors, no one could refute that the resemblances between the two are noticeably evident (see Table 2.7).

---

30 Why do orchestras need a conductor? Available at: https://www.classicfm.com/discover-music/instruments/conductor/what-does-a-conductor-actually-do/ [Accessed March 15, 2021].

	General Contractors▼	**Music Conductors▼**
Involved in	All types of construction projects, including large ones too (e.g. smart constructions)	Symphony orchestras and philharmonics, primarily large ones
Other given titles	Main contractor, primary contractor, prime contractor, and others	Chief conductor, main conductor, senior conductor, maestro, and others
An expert	Yes	Yes
A connector	Yes	Yes
Main duties	• Supervise and manage the execution of construction projects	• Ensure the accurateness of musical pieces and symphonies performed
	• Communicate with stakeholders at all times to keep them informed of progress and potential snags	• Communicate with musicians at all times during performances using hand gestures to keep them orchestrated
	• Have their own networks of construction actors	• Recruit instrumentalists
	• Conduct pilot tests and feasibility studies	• Run rehearsals and auditions
	• Direct and (co-direct) projects (scenarios A & B)	• Conduct (and co-conduct) concerts
	• Decide on construction project management processes	• Select entire music programs
	• Ensure fair interplay between all stakeholders	• Make sure all performers are playing harmonically
	• Handle non-technical tasks (e.g. scheduling, planning, following-up with clients)	• Handle managerial tasks (e.g. scheduling, planning, interacting with the audience)

	General Contractors ▼	Music Conductors ▼
Objectives	• Customer satisfaction	• Audience satisfaction
	• Deliver project on time and within budget	• Deliver great performance by setting the right tempo and shaping the sound of the ensemble
	• Bringing construction actors together to create an interlocked working system	• Unifying performers to create one sound, one orchestra

Table 2.7. *General contractors vs. music conductors (source: created by the author)*

2.6. Conclusion

Thus far, we have shown how the co*nstruction value chain* has evolved with the wave of *innovations in construction*, paving the way for the involvement of a new actor, the *central operator*. The latter is to act as a *conductor*, harmonizing the performance of all industry players so that large-scale, smart developments end up being impeccably built. Following Stokowski's Shift, we have also established that it is inconceivable that a single value chain can fit all types of projects; hence the need to conceive new ones (one or more) that could establish a better interplay between instrumentalists, including ICT companies and end-users. Similarly, we have evidenced that *business models*, which endure for longer than *construction project processes*, are highly influential when it comes to companies' long-term business success. Additionally, we have ascertained that GCs are both mavens and connectors in the field of construction, with their own support networks. Much like music conductors, GCs are frontrunners with the ability to play various roles and assume different positions across the construction industry's value chain: CMs, PMs, assistants to project owners and delegated project owners.

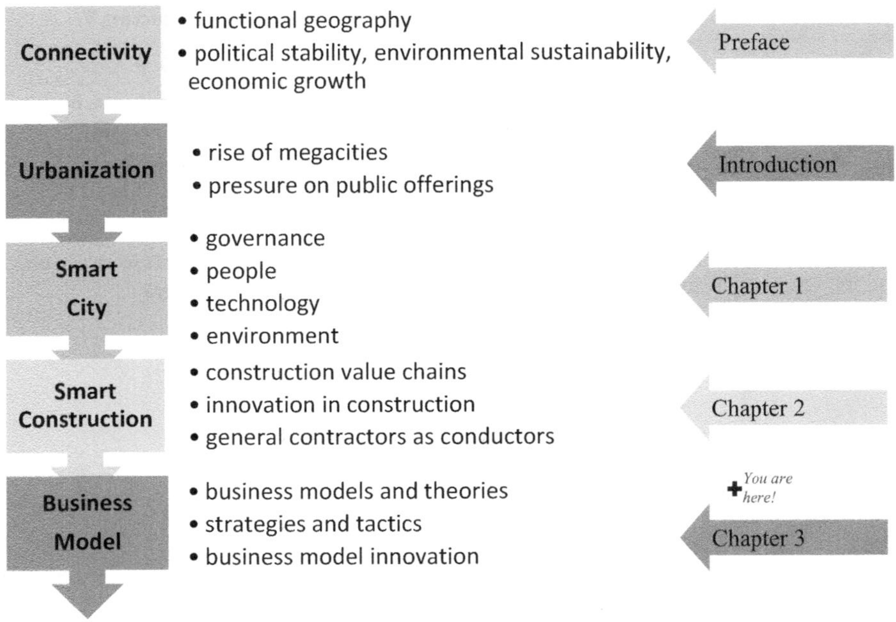
Connectivity
• functional geography
• political stability, environmental sustainability, economic growth
Preface
Urbanization
• rise of megacities
• pressure on public offerings
Introduction
Smart City
• governance
• people
• technology
• environment
Chapter 1
Smart Construction
• construction value chains
• innovation in construction
• general contractors as conductors
Chapter 2
Business Model
• business models and theories
• strategies and tactics
• business model innovation
You are here!
Chapter 3

3

On Business Models

This chapter examines the existing academic literature on *business models*. This is fundamental because the prime aim of this study is to design a new business model that would be a better fit, say more appropriate for the handling of large construction projects. The business model that we intend to build is that of a general contractor (*strategi* or central operator) who would be consigned (by a project owner) for the administration of smart developments, from conception until delivery of the final product. Yet, for this to be conceivable, we, to begin, endeavor to comprehend the specificities of business models. We will then search to recognize the various paths to business model innovation. In other words, we intend to describe the rationales based on which companies seek to create and capture value in various settings (economic, social and environmental) and elucidate the process of business model innovation which, we believe, is part of a company's open set of business strategies. With that said, Chapter 3 is structured as follows:

– First, the business model notion is clarified – an overview of the most prominent representations of business models is presented – and the relationship between different sets of business model arrangements and value creation is embarked upon.

– Next, we demonstrate how business models are forthrightly linked to a company's business strategies and, as a result, are decisive regarding its business tactics.

– Then, we examine the process of business model innovation, as a means for crafting, applying and authenticating new business models. Also, we pinpoint the major gaps in the research on business model innovation today.

– Finally, we provide some real-world business model innovation examples. By discussing the best managerial practices that could help executives in general, specifically those in the construction industry, we take business model innovation to the level of a consistent and corrigible discipline.

3.1. Definitions

Almost 30 years ago, following the rise of the World Wide Web in the mid-nineties, the BM (Business Model) concept has been envisaged by numerous authors via distinct subject-matter lenses, and has been gaining ground since then (Zott *et al.* 2011). It is only after the advent of computers and spreadsheets that BMs became popular (Magretta 2002). Prior to that, successful BMs were created unintentionally – neither by design nor by foresight – and became clear only after the fact. At the outset, BMs were only concerned with e-business (e.g. e-commerce and e-market) and it was only after a short while that the scope of BMs became protracted (Timmers 1998) and BMs were used to assess other phenomena (e.g. innovation and technology management). Indeed, BMs came to be the new basis of competition, replacing product features and benefits as the playing field on which companies emerge as dominant or laggards (Chesbrough and Rosenbloom 2002). Nowadays, the existing literature on BMs suggests the prevalence of four school of thought[1] (Zott *et al.* 2011), explicitly: (1) *BM* as a *central unit of analysis*; (2) *BM* as a *system-level, holistic approach* to expounding how companies capture value and generate profit; (3) *BM* as a *strategic means* to explain both value creation and value capture phenomena; and (4) *BM* as a *platform* where the activities of focal companies and their partners are expected to play a key role in the conceptualization of BMs in force.

Moving on, we wonder: *What exactly is a BM*? In spite of the recent deluge in the literature on BMs, we affirm that a definitional clarity on BMs is still nonexistent today (Zott *et al.* 2011; Bucherer *et al.* 2012), and BMs are repeatedly studied without a clear definition of their concept (we remain optimistic nonetheless, trusting that these emerging themes may one day lead to a more fused study of BMs). According to Massa *et al.* (2017), BMs constitute a key vehicle for innovation at the corporate level – and this, in two ways: (1) by allowing executives to connect innovative products and technologies to a realized output in a market; and (2) by being a distinct source of innovation *per se*, paired to customary dimensions of innovation (product, process, managerial and others). Likewise, Lambert and Davidson (2013) and DaSilva and Trkman (2014) acknowledged that a BM exhibits how a company generates, conveys and captures value. It underlines the content, configuration and management of the activity system that carries a value proposition to the customer.

For Ferrero *et al.* (2015), BMs are the method that a company uses for its functioning – encompassing the aim, schemes and persons who join forces to bring

1 In contrast, other authors such as Gassman *et al.* (2014) talked about seven schools of thought, yet for the purpose of this book, we stick to the classification suggested by Zott *et al.* (2011) and signpost that there is still no common definition for BMs among scholars.

added value to customers. A simple case of customer needs being fulfilled by totally distinct BMs, the authors specified, is *car transportation*. Getting from point A to point B could be served by a car that is either bought, leased, rented or shared. The main product is basically the same, while the activity system underlying the product shields a spectrum of distinct BM design choices. Overall, value capture is enabled by the company's ability to homogenize and mechanize operations across the activity system – and the purpose is not to add intricacy, but to allow for scalability via simple, operative activity systems.

Going forward, Micheal Lewis in his book entitled *The New, New Thing: A Silicon Valley Story*[2] spoke of BMs as *artworks*. And like any artistic production, he emphasized, it is one of those things that people sense they could spot at first sight without being able to explain them. Indeed, he suggested the meekest of definitions: "it is how a company plans to make money". By the same token, Magretta (2002) avowed that BMs are stories – stories that elucidate how companies work. The author declared, "any decent BM would be able to answer the following age-old questions: Who is the customer? What does the customer value? How does the business work? And how could value be delivered to customers at an appropriate cost?" In the same context, Weill and Vitale (2001) recognized that BMs depict roles and rapports among a company's customers, partners, suppliers and manufacturers. Similarly, Morris *et al.* (2005) associated BMs with the manner – or the art – of conducting business to ensure the sustainability of the company, providing the following definition: "a BM is a method of doing business by which a company could sustain itself by generating revenues". "A BM tells how a company makes money by identifying where it is positioned in the value chain", the authors added.

As for Amit and Zott (2001), a BM describes the content, structure and governance of transactions designed to create value through the exploitation of business opportunities. A BM, they added, is conceived by a company to create and capture the largest market value possible from the products and/or technologies it is commercializing. The quality and the level of the economic outcomes, they concluded, depend on the efficiency of the BM being used. Subsequently, they asserted, "the more a BM is efficient and ingenious, the more the market value of the newly commercialized products, created and captured through it, would be great". Osterwalder (2004), we proclaim, had a significant contribution in this respect too – defining a BM as a conceptual tool that contains a set of interconnected elements – and allows one to express a company's logic of making money. In a similar manner, Demil and Lecocq (2010) demarcated a BM as the way a company

2 A soft copy of the book is available at: https://epdf.pub/the-new-new-thing-a-silicon-valley-story.html.

operates to ensure its sustainability: "Companies should focus on sources of value creation that are inherent in the BM itself and could be accessed through it".

On another note, Zott and Amit (2010) signposted that BM design is a key decision for start-ups – and a critical, perhaps more difficult one for established companies with prospective plans to redefine their existing business structures. The authors pictured a BM as an *activity-system-design framework* that encourages holistic thinking (see Table 3.1). As reported, "a BM is geared towards total value creation for all parties; the greater the total, the greater the company's bargaining power, and the greater the amount of value it could appropriate; it is a template of how a company conducts business and how it delivers value to its stakeholders."

Design Elements	
Content	What activities should be performed?
Structure	How should they be linked and sequenced?
Governance	Who should perform them, and where?
Design Themes	
Novelty	Adopt innovative content, structure or governance
Lock-In	Build in elements to retain business model stakeholders
Complementarities	Bundle activities to generate more value
Efficiency	Reorganize activities to reduce transaction costs

Table 3.1. *An activity system design framework (source: adapted from Zott and Amit (2010))*

A fairly comparable definition can be found in Demil *et al.* (2015): "a BM enlightens the logic of a company, the way it operates, and how it creates and captures value for its stakeholders." Table 3.2 lists some other prevalent definitions suggested for BMs.

To finish, we say, given the scope of our book and the objectives assigned to it, we choose to adopt Timmer's (1998) definition of BMs; also, to some extent that of Demil *et al.* (2015). As we see it, BMs are meek strategic tools that companies resort to, to better depict the relationships they hold with different actors (suppliers, customers, etc.) along the industry value chain. These tools, we affirm, enable companies to better visualize the practices through which they capture and create value. From our perspective (as an expert in the French construction industry), a BM is a give-and-take, looping process that describes a company's moneymaking rationale.

Definition	*Author/s*
The BM is an architecture of the product, service and information flows, including a description of the various business actors and their roles; a description of the potential benefits for the various business actors; a description of the sources of revenues.	Timmers (1998)
The BM is the heuristic logic that connects technical potential with the realization of economic value.	Chesbrough and Rosenbloom (2002)
BMs consist of four interlocking elements, which, taken together, create and deliver value. These are customer value proposition, profit formula, key resources, and key processes.	Johnson *et al.* (2008)
A BM is a *reflection* of the company's *realized* strategy.	Casadesus-Masanell and Ricart (2010)
A BM articulates the logic, the data and other evidence that support a value proposition for the customer, and a viable structure of revenues and costs for the company delivering that value.	Teece (2010)
A BM is a meta concept to exemplify the business strategy of a company. There are four components of a BM: customer, customer engagement, monetization and value chain, and linking mechanisms — which aim at capturing the essence of the cause-effect rapports between customers, the company and money.	Baden-Fuller and Mangematin (2013)

Table 3.2. *BM definitions (source: compiled from various sources)*

The next section delves into the main constituents and representations of BMs – exactly as found in the literature stream on BMs.

3.2. Constituents and illustrations

The Internet was the prime driver that led to the growth of interest in BMs – and the resulting rise of an extensive literature that spins around the topic (Magretta 2002; Yip 2004). In fact, the number of e-business scholars who aimed to differentiate between primary- and secondary-class[3] areas of BMs' constituents has significantly grown over time. A listing of those is available in Table 3.3.

3 Denoting respectively most and (relatively) less important business models' components.

Primary-class areas	*Secondary-class areas*	*Author/s*
• Value stream for partners and buyers network • Revenue stream • Logistical stream	n/a	Mahadevan (2000)
• Profit stream	• Customer selection • Value capture • Differentiation and strategic control • Scope	Stewart and Zhao (2000)
• A system made of components, linkages between components, and dynamics • Customer value • Revenue sources	• Scope • Price • Connected activities • Implementation • Capabilities • Sustainability	Afuah and Tucci (2001)
• Mission • Structure • Processes • Revenues • Legal issues • Technology	• *Mission*: goals, vision, value proposition • *Structure*: actors and governance, focus • *Processes*: customer orientation, coordination mechanism • *Revenues*: source, business logic	Alt and Zimmerman (2001)
• Concept • Capabilities • Value	• *Concept*: market opportunity, product and service offered, competitive dynamics, strategy for capturing a dominant position, strategic options for evolving the business • *Capabilities*: people and partners, organization and culture, operating model, marketing sales model, management model, business development model, infrastructure model • *Value*: benefits returned to stakeholders, benefits returned to the company, market share and performance, brand and reputation, financial performance	Applegate (2001)

Primary-class areas	*Secondary-class areas*	*Author/s*
• Sustainability • Revenue stream • Cost structure • Value chain positioning	n/a	Rappa (2001)
• Value proposition • Customer segments • Partners' network • Delivery channel • Revenue stream	• Relationship • Value configuration • Capability • Cost structure	Osterwalder (2004)
• Products and services delivery • Customers • Costs structure • Income	• Network • Network externalities	Bonaccorsi *et al.* (2006)
• Costs • Revenue stream • Sustainable income generation • Goods and services production and exchanges	• Pricing strategies • Relationships • Network externalities	Brousseau and Penard (2006)

Table 3.3. *The constituents of business models (source: adapted from Zott* et al. *(2011))*

To make a long story short, scholars (Amit and Zott 2001; Lucking-Reiley and Spulber 2001; Afuah 2004) who concentrated on e-business as an area of research on BMs have been interested in understanding the functioning and operability of companies engaging in unusual, Internet-based ways of doing business – as well as in the first-hand roles that those companies could play in their respective ecosystems. To do so, they have invested lots of time and effort in defining and representing generic BMs, rather than propose causal justifications or conduct empirical analyses. Their descriptive contributions, we assert, highlighted, to some extent, the notion of value (customer value, value proposition and others), financial aspects (e.g. revenue streams and cost structures) and other elements associated with the architecture of the network between the company and its partners (delivery channels, network relationships and others).

As per Zott *et al.* (2011), each of these constituents are possibly to account for part of a generic BM, and could help scholars set different arrangements (types) of

BMs apart. In fact, the authors added, a BM is neither a value proposition, a revenue model, nor a network of relationships *per se* – but instead is all of these elements combined.

Up until now, many academic researchers have tried to embody BMs via a blend of informal textual and/or purposeful graphical illustrations. Weill and Vitale (2001), for instance, presented a set of simple graphics envisioned to provide tools for the study and design of e-business initiatives. The graphics of their BM focused on three classes of matters: *participants*, *rapports* and *flows*.

Equally importantly, Tapscott *et al.* (2000) suggested a *value mapping approach* to portray how a business web actually works. The latter shows all key classes of *participants* and *value exchanges* between them – and exhibits the shape and nature of a BM, thus enabling companies to run plausible BM simulations before engaging in real-life projects. Other scholars have provided a *BM ontology* – being a conceptualization and validation of the elements, rapports and lexis of a BM (Osterwalder 2004), organized into several stages of decomposition with snowballing levels of complexity. In the same manner, Gordijn and Akkermans (2001) came up with a *conceptual modeling approach* for the representation of BMs. Their ontology scrounges notions from the business literature (actors, value exchanges and activities, and others) – and uses them to model networked assemblages of companies and end-users who create, dispense and consume things of economic value.

The upcoming sections list and explain four noticeable BM representations as well as clarify how those representations (not all, but some of them!) serve the purpose of our empirical research in this book – and to what extent.

3.2.1. *Business model canvas*

Osterwalder's (2004) conceptual tool encloses a set of interrelated elements, which set the path to the clear-cut expression of a company's BM. As indicated by the author, "it is implausible for companies to be able to develop or design innovative and effective BMs unless they pay thorough attention to their constituents". The competent conception of a new BM – whether a company intends to redesign or reinvent an existing BM, or even create an entirely new one – is indisputably a tricky event (Gassman *et al.* 2014).

Overall, companies often use Osterwalder's BM canvas as aid to develop new BMs – or to help analyze and map existing ones. Originally, the latter was inspired by the works of various scholars (Afuah and Tucci 2001; Amit and Zott 2001; Applegate 2001; Weill and Vitale 2001; Chesbrough and Rosenbloom 2002;

Magretta 2002); specifically, Osterwalder (2004) identified the building blocks of existing BMs – as exhibited in the literature, then synthetized and conceptualized them, and then put them together to create one puzzle, a distinctive BM (Figure 3.1).

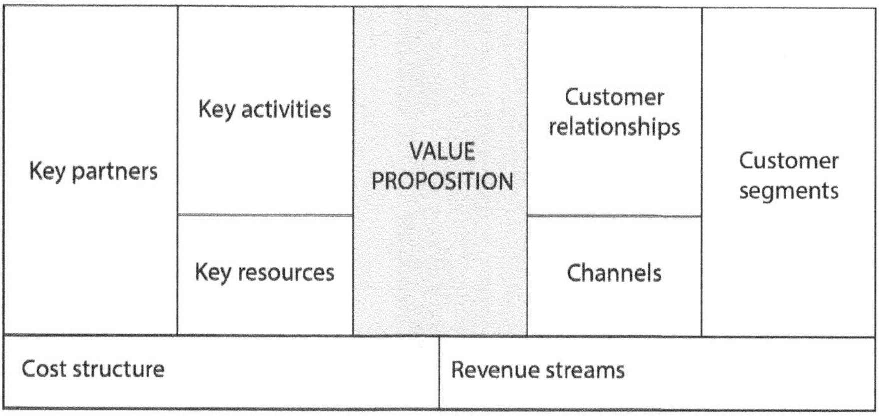

Figure 3.1. *Osterwalder's business model canvas (source: adapted from Osterwalder (2004))*

If we focus in on the BM under examination, we get a vision of the latter's ontology (Figure 3.2), that is, a detailed sight of the various elements and rapports describing the money-earning logic of a company. Largely, the ontology comprises *nine building blocks*, the so-called *BM elements*, where each *element* (e.g. the *value proposition*) could be broken down into a distinctive *sub-element* (e.g. the *offering*). This breakdown allows for the reconsideration of BMs on different levels of granularity and based on particular needs. Indeed, in order to better comprehend value and to construct innovative offerings, Osterwalder (2004) suggested a *conceptual (systematic) approach* outlined in the value proposition element that makes value innovation easier. This, according to the author, would enable companies to identify and map their existing value proposition and compare it to that of their rivals. Osterwalder's BM, we stress, revolves around one dominant notion: *value proposition*. It is the first and most central of the nine elements of BM ontology, perceived as the declarations of remunerations that are transported by the company to its external populations (Bagchi and Tulskie 2000). Plainly, *value proposition* denotes how items of value are swathed and offered to satisfy customer needs; it describes the way a company sets itself apart from the competition and is the main reason why customers choose to deal with one company instead of another[4].

4 Ibid.

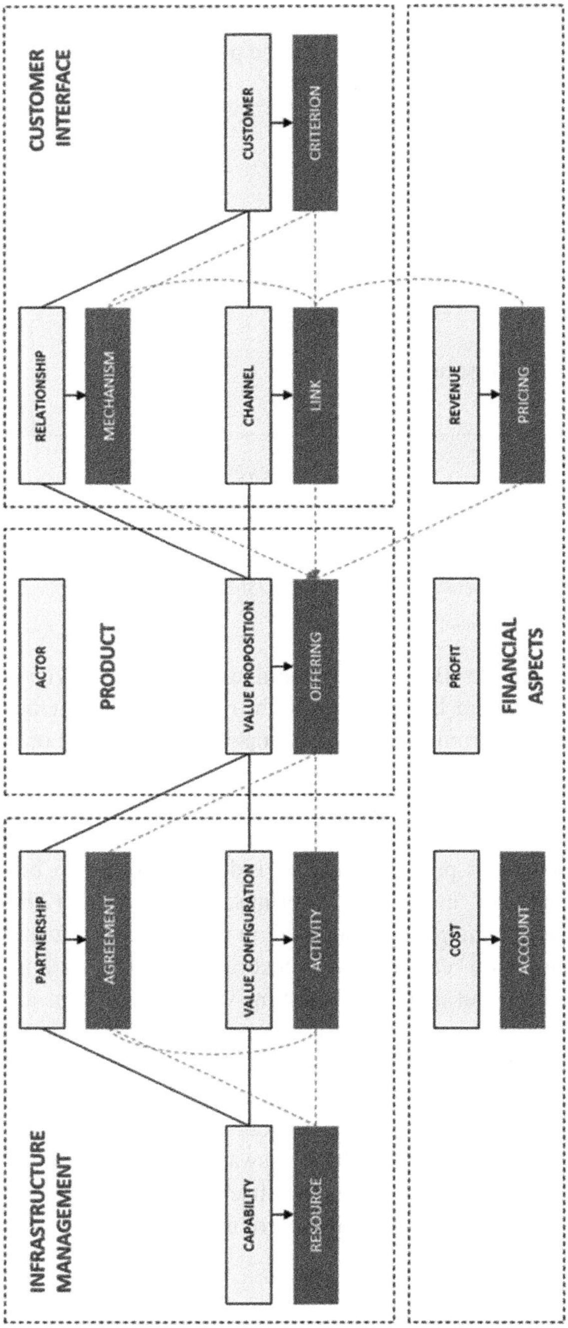

Figure 3.2. *Osterwalder's business model ontology (source: adapted from Osterwalder (2004) and Osterwalder et al. (2005)). For a color version of this figure, see www.iste.co.uk/karam/general.zip*

By itself, Osterwalder's BM canvas is global and comprehensive, at least from an economic standpoint. This is accurate as it covers all BM elements spread along the value chain, *up-* and *down-stream* – going from the company's dealings with partners to production activities, bearing in mind the availability of resources (*infrastructure management*) – and then protracting to shield the logistics needed for the commercialization of products or services to targeted market and customer segments (*customer interface*). Also, it helps a company ensure a better trade-off between its cost structures, on the one hand, and its revenue streams, on the other hand (*financial aspects*). The limit of the model, nevertheless, lies in the fact that it is *profit-oriented*, solely focusing on the economic aspect of business activities, hence overlooking some other, equally important aspects, for example, those that are *social* and *environmental*.

3.2.2. *Causal loop diagram*

Casadesus-Masanell and Ricart (2007) indicated that developing new BMs would enable companies to strengthen and sustain their current market positions by developing and implementing new products, services or even technologies. The authors praised the *inductive perspective* to building BMs in general – through a CLD (Causal Loop Diagram)[5]. "A BM consists of different part sets", the authors professed, "and the choices made on how the company should operate – and the prospective consequences that might ensue from these choices are interrelated [...] which infers that every corporate choice made would inevitably lead to some consequences, one or more, positive or negative". Indeed, as put by Kiani *et al.* (2009), CLD is a valuable tool that helps companies capture the gestalt of BMs and provide executives with significant insight into BMs to better apply them. BMs, they said, show how companies operate by choosing how to liaise with their adjacent environments.

Looking at Figure 3.3, Casadesus-Masanell and Ricart (2011) divulged that corporate decisions are threefold. First comes *policy choices*, which slot in all means of action that a company employs to cover the different facets of its operations. *Asset choices* follows, invoking decisions made vis-à-vis tangible resources. Third and last, there are the *governance choices*, which designate the structure of contractual arrangements that confer decision rights over policies and assets. In order to better elucidate the imbrications of choices versus consequences, the authors gave the example of the BM of a company that operates a fleet of trucks (*asset choice*). The company, they said, has a *governance choice* to make between buying and owning the trucks, or leasing and operating them. No matter the

5 A Causal Loop Diagram helps with understanding and analyzing complex systems by identifying the key variables in a system, and the cause and effect relationship between them.

company's choice, they added, the decision made, true or false, right or wrong, would potentially pointedly sway its ability to create and capture value, positively or negatively.

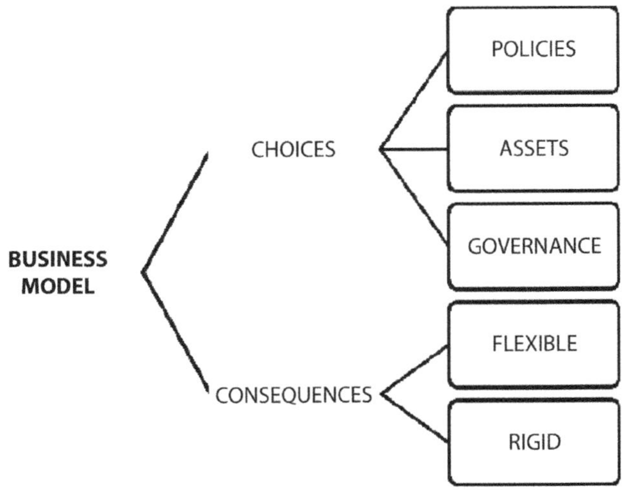

Figure 3.3. *Elements of a business model (source: adapted from Casadesus-Masanell and Ricart (2011))*

Choices	*Consequences*
Secondary airports	↓ Airport fees
↓ Tickets prices	↑ Volume
↓ Commissions to travel agents	↓ Cost
Standardized fleet of 737s	Bargaining power with suppliers
Single-class	Economies of scale
High-powered incentives	Attracts combative team
No meals	Faster turnaround
Nothing free	↑ Revenue
Spartan headquarters	↓ Fixed cost
No unions	Flexibility in rostering staff

Table 3.4. *Ryanair: choices and consequences (source: adapted from Casadesus-Masanell and Ricart (2011))*

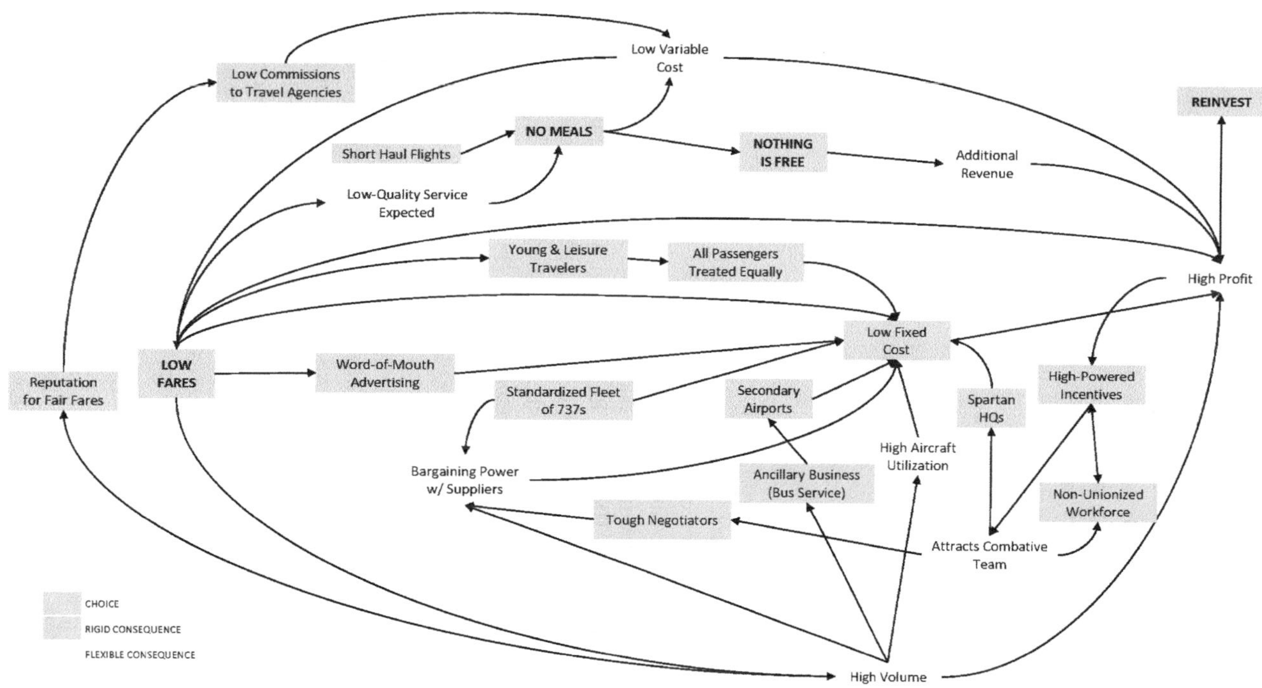

Figure 3.4. *Ryanair's (low-cost strategy) business model (source: adapted from Casadesus-Masanell and Ricart (2011)). For a color version of this figure, see www.iste.co.uk/karam/general.zip*

Ryanair[6] is another great real-world example that exhibits how a company could use CLD to depict its own operating BM. Table 3.4 summarizes the choices made (or the business strategies opted for) by Ryanair versus the consequences that later ensued as a result of these choices – which eventually resulted in the creation of the so-called Ryanair BM – also known as Ryanair's low-cost strategy BM. According to Casadesus-Masanell and Ricart (2010, 2011), it is the recurrent process of dotting the causality linkages between various choices and consequences that would convert a company's BM into a CLD (Figure 3.4). "The choices made and the suing consequences are creators of BMs", the authors added. "And the rigidity or flexibility of a given consequence is measured by its elasticity or inelasticity to the variations in the choices made."

Before we jump on to explaining the RCOV framework, we end this section by mentioning that the logic behind the CLD is partially espoused in our empirical research to build and conceptualize the *stage-based construction process map*, a *piloting tool* by which we mark all the potential connections among and within the elements of our envisioned BM.

3.2.3. *RCOV framework*

Similarly to other scholars, Demil and Lecocq (2010) contemplated BMs from a *configurational viewpoint*, and came up with the RCOV framework (Figure 3.5) which is currently one of the most cited representations in the academic literature on BMs. Because of its dynamism, features and parsimony, the latter seems to have a certain edge over other comparable BMs. The RCOV model was first known as the RCOA model (Lecocq *et al.* 2006; Demil and Lecocq 2008) but as a result of subsequent refinement, it took a new form and was therefore labeled otherwise. RCOV stands for *Resources & Competences* (RC), *Organization* (O) and *Value proposition* (V), denoting the three basic elements of the BM under examination, whereas RCOA is short for *Resources & Competences* (RC), *Offerings* (O) and *Activities* (A).

In a broad sense, the RCOV model follows a defined logic that could be precised in the following manner: a company builds its BM by making various choices to generate revenue from different sources. And the choices made comprise a chain of *resources and competences* to be valued, namely: a business-to-customer *value proposition* and the business *organization per se*, internal and external. Moving forward, the *resources and competences* are valued through the supply of products or services on markets. The *organization*, on the other hand, refers to the set of operations that a company deals with and chooses from (*value chain*) and the associations it holds with other stakeholders or partners (*value network*).

6 Ryanair is a low-cost airline, focused on offering cheap fares in order to capitalize on the market; it offered fares, which were 20% lower than the cheapest fare of its rivals. Also, through the standardization and diversification of its services, the company was able to uphold, even boost its standing in the marketplace.

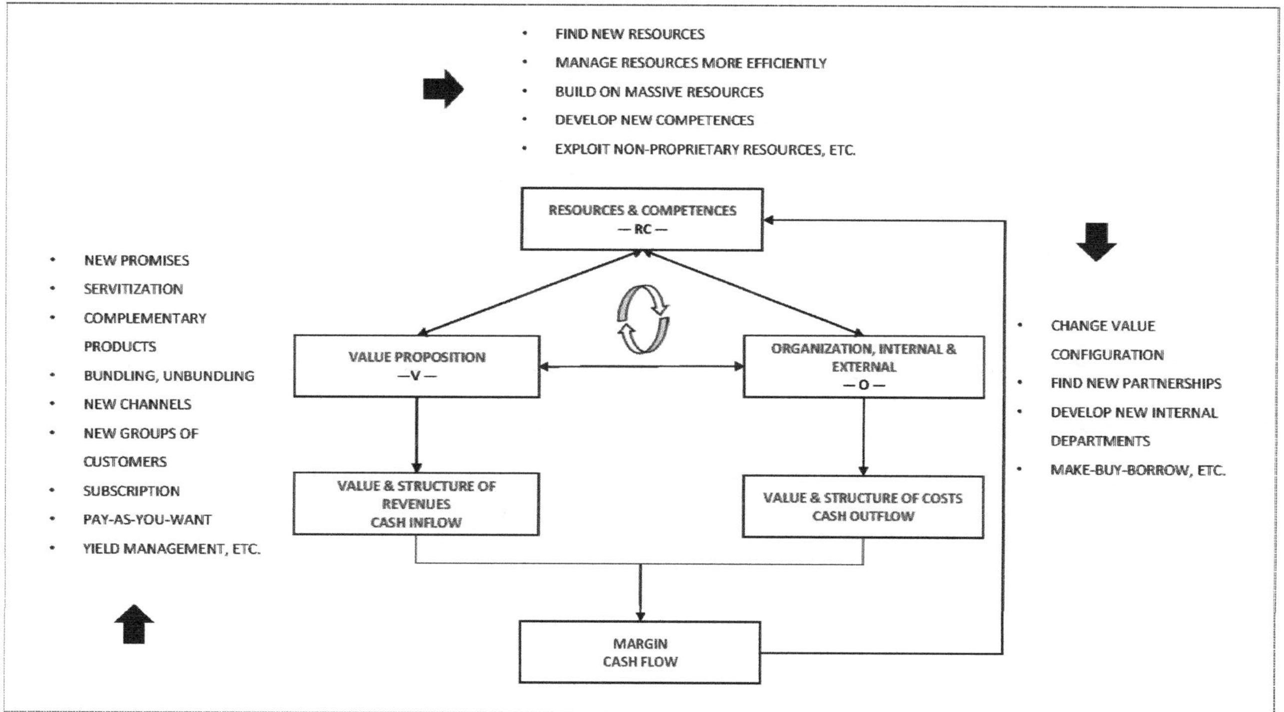

Figure 3.5. *RCOV framework (source: adapted from Demil and Lecocq (2010)). For a color version of this figure, see www.iste.co.uk/karam/general.zip*

The model's three basic elements determine the structure and the size of costs and revenues of a business and therefore its profit margin, as well as the sustainability of its BM. The *cost structure* is driven by a company's organizational configuration, while the *revenue structure* is based on the exclusivity and appeal of the products or services it offers to its customers.

To summarize, we note that, while the customer is not a direct part of the RCOV model, it is – or at least may be – a constituent of a company's external managerial structure, thus part of one of its main elements (*Organization*). This is important because customers are certain to play a decisive role in the success (or failure) of a company's BM. Thus, incorporating *customers* into a company's BM is crucial, both in theoretical and practical terms. A comparable proclamation was repeatedly made throughout this book (see Chapter 1) to underline the pivotal role that people (a primary smart city component) could play (as producers of data and value) in the conception and design of successful smart cities. Moreover, based on the RCOV framework, it is sort of palpable that the positioning of a given company within the value network is critical for its long-term business success; also, the way the latter chooses to integrate both *Resources & Competences* components in its own BM would eventually sway its organizational structure and influence its ability to create and capture value:

Recapping, the TLBMC (triple layered business model canvas) constitutes the foundation of our envisioned solution-based BM, which aims at identifying the main problems faced at the construction industry level – as well as at suggesting actionable insights as to what the possible solutions might be. The CLD on the other hand informs the carcass of our piloting tool. Through dependency graphs, the latter would enable construction companies to track their works every step of the way. As for the RCOV framework, we take the rationale behind it, especially its R-C-O components, to further elucidate the budding changes in our new BM.

Now that we have defined BMs, demarcated their integral components, and explained their conceptual representations, the next section identifies different *business model arrangements* that describe and clarify the various ways companies engage with stakeholders to create and capture value.

3.2.4. *Triple layered business model canvas*

Pigneur *et al.* (2015) expanded on Osterwalder's BM canvas – which consisted of only one layer (*economic*) – by adding two extra layers to it: *environmental* and *social*, thus rendering it more *sustainable* and *customer-integrated* (Figure 3.6).

As per Sherman (2012), these supplementary layers parallel Osterwalder's BM canvas by underlining the interconnections, which support environmental and social impacts disjointedly, and lengthen it by drawing links across all three layers to bear a cohesive triple bottom line perspective of managerial impact. The TLBMC (triple layered business model canvas) is one of the best strategic management tools that companies can use to explore the sustainability of their new BMs, as it explicitly shows how they could bring value to their businesses (Abraham 2013). The said BM representation is interesting by itself because it did well by converting a *profit-oriented* BM (Osterwalder's original representation) into a *human-centric* one. At its core, the TLBMC promotes a shift from incremental and secluded innovations towards more integrated and complete sustainability-oriented innovations, which, we trust, are better suited to meeting ongoing international crises, and energy and factor constraints (Adams *et al.* 2015).

As a tool, the TLBMC bridges business model innovation (Spieth *et al.* 2014) and sustainable BM development (Boons and Lüdeke-Freund 2013), and helps companies overcome barriers to sustainability-oriented change (Shrivastava and Statler 2012; Lozano 2013) by allowing them to creatively rethink their current BMs and communicate potential innovations.

Furthermore, as a multi-layer BM canvas, the TLBMC endows companies with a clear and somewhat easy way to visualize and discuss the multiple and diverse impacts of their BMs. Therefore, instead of attempting to condense multiple types of values into a single canvas, the TLBMC allows companies to explore economic, environmental and social values in a bidirectional manner, horizontally and vertically, among and within layers (and building blocks). Simply put, the TLBMC provides a concise framework to encourage visualization, communication and collaboration around innovating more sustainable BMs (Boons and Lüdeke-Freund 2013). In contrast to Osterwalder's BM canvas, the TLBMC offers companies the opportunity to address a triple bottom line where each canvas layer is dedicated to a single dimension, and together they provide a means to integrate the associations and impacts across layers.

Putting the *economic layer* aside, the *environmental layer* of the TLBMC, as mentioned by Abu-Jbara (2019), stems from research and practice on LCA (Lifecycle Assessments), which offers an assessment of environmental impacts across multiple types of indicators over the entire lifecycle of a product. Linking LCA to business innovation may possibly support competitive product and business model innovation, and hearten continuing impact measurement and improvement of sustainability-oriented innovations over time (Chun and Lee 2013). With regard to the TLBMC's *social layer*, it strives to balance the interests of a company's stakeholders rather than just focus on profit maximization. Much like the *environmental layer*, the *social layer* extends the original BM canvas by scrutinizing a company's BM based on a *stakeholder perspective* (Mitchell and Coles 2003). Upon reexamining Figure 3.3, we recognize that the TLBMC's *environmental layer* revolves around one chief element, the *functional value*, which defines the focal output of a product or service delivered by a company. The *social value*, on the other hand, is the TLBMC's *social layer*'s chief element, and relates to a company's mission which focuses on creating advantages for both its stakeholders and the society as a whole.

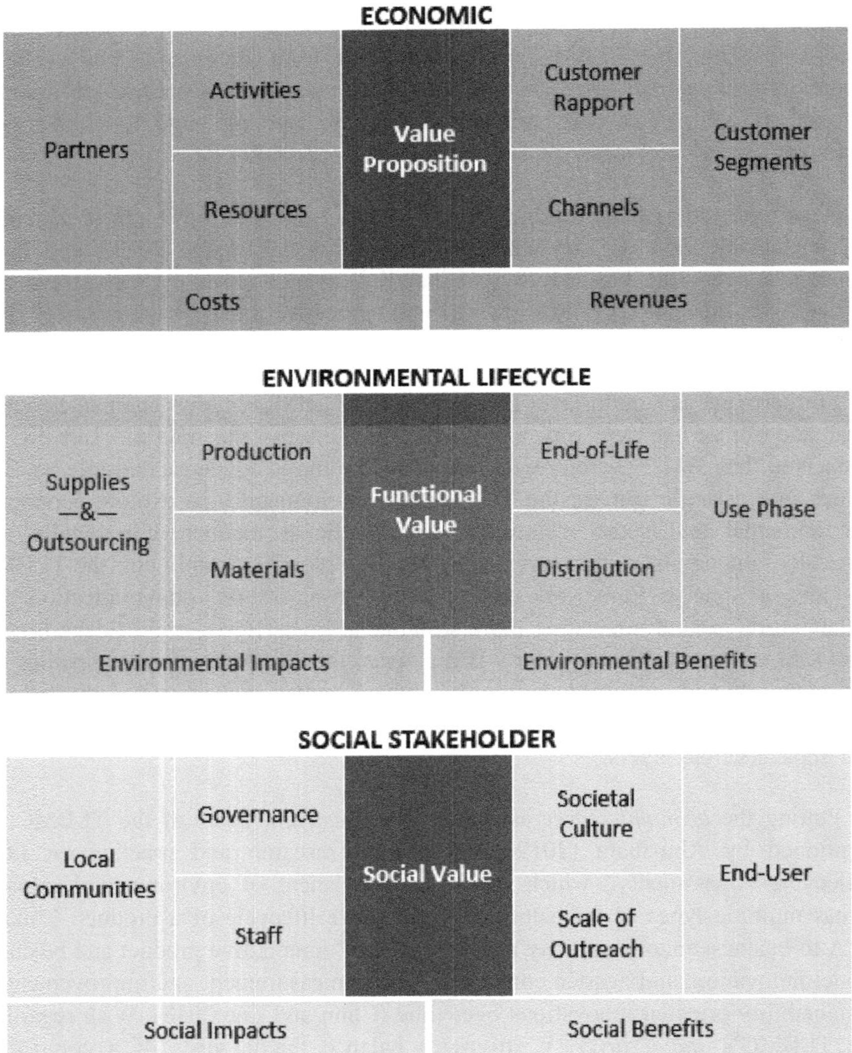

Figure 3.6. *The triple layered business model canvas (source: adapted from Pigneur et al. (2015)). For a color version of this figure, see www.iste.co.uk/karam/general.zip*

As far as we are concerned, for the purpose of this book, we choose to adopt the TLBMC as a means or strategic tool for conducting BM analysis and innovating towards a more sustainable, human-centric BM in construction (that of a *general contractor* for the managing of smart developments). The TLBMC, we believe, constitutes the foundation for our envisioned BM, which accounts for smart city components (*people*, *living*, *governance*, *economy*, *mobility* and *environment*). The similarities between smart city components (on the one hand), and the TLBMC's layers and building blocks (on the other hand) are obvious, even at first sight; hence our choice of BM canvas (continue reading for more details in this regard).

3.3. Business model arrangements and value

This section builds on various insights on customer[7] value to identify the different sets of arrangements (types) depicted in BMs that describe how companies prompt *value creation* (for customers) through *value networking* (with partners). The most relevant filament of the literature that links with our research goal is the *conceptual approach*, which gives *value* a chief role (Teece 2010) and sees the BM as a hasty and simple illustration (see previous sections) of how companies generate financial earnings (Chesbrough 2010; Baden-Fuller and Haefliger 2013).

As formerly noted, the BM literature which was, customarily, to some extent, regardless of the reasons, careless about customer value (Pauwels and Weiss 2008; Osterwalder and Pigneur 2010; Zott and Amit 2010; Zott *et al.* 2011) has now become (unsurprisingly!) more and more customer-oriented (Baden-Fuller and Haefliger 2013):

> The evolution of the smart city concept followed a somehow similar path. After the advent of the World Wide Web, the modern city archetype has become a mere technology demonstrator – and it is only recently that it became people-centered, conceived, and administered following a bottom-up approach.

The main idea conversed herein borrows from the marketing literature (Vargo and Lusch 2004; Gronroos and Voima 2013), spins around the fact that customers have become *co-producers of value*, and comes hand in hand with the sociological view that value creation necessitates *social construction* (Pinch and Bijker 1984):

> These assertions fit perfectly with the rationale behind our envisioned BM, where end-users – as creators of innovation – are expected to

7 The terms *client*, *customer*, *consumer*, *user* and *end-user* are used interchangeably throughout this section.

play a vital role in the building of smart cities and the handling of smart construction projects.

We take these ideas as given, and explore the first pair of BM *dyadic* arrangements (Baden-Fuller *et al.* 2017) for the co-creation of value: *product BM* versus *solution BM*. Yet, before we delve into an explanation of these, we reasonably start by settling on what *value* is. For this research, we assume *value* to denote *use-value*, standing for the *utility* that customers may gain from using a particular product or service. This is because, as put by Vargo and Lusch (2004), a product or service is considered worthless unless the customer could (literally!) use it or do something beneficial with it, for enjoyment or any other purpose.

The PBM (Product Business Model): when use-value activities are performed, succeeding the act of purchase, and away from the provider. In such a case, the *providers*' arrangements for engaging with *customers* are labeled the *PBM* (Baden-Fuller *et al.* 2017). Even though this conception of the product and its use-value is captured in the sphere of the *final customer* only, the exact same logic could also apply to a *middle agent* representing the final customer – such as in business-to-business situations where an *actor* in the value chain may step on the scene on behalf of the *end-user*.

The SBM (Solution Business Model): when the *provider* directly engages the *final customer* in the production process, resulting in the latter's use-value being co-created. This situation is labeled the *SBM* (Gronroos and Voima 2013; Baden-Fuller *et al.* 2017). From the standpoint of the *provider*, the skills and assets needed to produce a *solution* are not the same – being far greater than those required by a *product*. In such a setting, the *end-user* is more involved in the process and sometimes forced to change their demeanor to fit new situations. We add that an *SBM* is said to be meaningful to the *end-user* – but if and only if the ensuing use-value is larger than that gotten from the *PBM*:

Our envisioned general contractor BM is human-centric, based on the active involvement of end-users throughout the whole project management construction process, as data providers and creators of innovation. Unmistakably, the PBM relates to the old construction value chain where construction companies tend to care more about the expected value of their profit margins and less about the well-being of their customers and the quality of their deliverables. In contrast, the SBM bears a resemblance to the new construction value chain, involving a central operator who will oversee the resourceful running of construction projects and ensure end-users are engaged every step of the way.

In the first pair of BM arrangements (the product- and solution-based BMs), there are only two players involved in the process of value creation. However, in certain cases, when the product is far more complex, there might be more than just one provider on stage, thus transforming the *dyadic* rapport into a *triadic* one. This is mainly the case of *paired providers*, selling *complementary products* (printers on one side, and paper and ink on the other). In such a case, the essential *dyadic*, two-player rapport remains unchanged – because only one *final customer* would ultimately benefit from use-value when all the elements have been put together (Teece 2010).

Our distinction between *dyadic* and *triadic* BMs echoes the literature on technology strategy – particularly the contributions of Zott and Amit (2010) and Thompson and MacMillan (2010), who disclosed that companies could have access to several value networks – meaning that technology has endowed companies with new forms of value creation gears, which are orchestrated in the sense that value is currently being created together, by a *company* and a surfeit of *partners* – for multiple *end-users*.

Now sliding to the *triadic* BM arrangements, we note that there are two of them (Baden-Fuller *et al.* 2017): the *matchmaking BM* and the *multisided BM*, explained below.

***The MMBM (Matchmaking Business Model)*:** involves three players: the *company* that arranges the market (the platform owner) and *two customer segments*, *buyers* and *sellers* who trade an underlying product or service. If we take the case of *Uber* for instance; *Uber*, being the *matchmaker*, would be behind *value creation*. For *buyers* (the persons looking for rides), the condensed search time and effort to find a ride means that they may enjoy greater use-value from the service provided via the *matchmaker*. For *Uber*, this increase in use-value is a source of potential yield. As for the *sellers*, they are paid on a ride-by-ride basis – therefore, the more rides they achieve (the more *buyers* they serve), the bigger their pays will be.

***The MSBM (Multisided Business Model)*:** entails three or more players, where the *company* (the platform owner) creates a set of dealings among *two or more otherwise detached customer segments*. The *first customer segment* is called the *customer beneficiary*; they receive a set of products or services at a trifling price, paid for by a *second segment* called the *paying-customer* who profits from creating value for the consumption of the first segment. Central to this BM is the win-win process by which the *two customer segments* are brought together, as follows: the *paying customer* profits from the *customer beneficiary*, using the product or service, and the *customer beneficiary* profits from the presence of the *paying customer*, paying for the product or service. Specialized magazines offering the option of free-of-charge advertisement constitute a great example of an *MSBM* arrangement (see Table 3.5 shown below for a quick overview of all four BM arrangements).

We finish by affirming that the list of BM arrangements exposed herein is extensive, yet non-exhaustive; also, we point out that technology has no objective value by itself and remains latent until it is commercialized via a competent BM (Chesbrough 2010). Indeed, according to the author, BM change is not solely driven by technology, but also by how we think and why we think that way; therefore, in most cases, managerial processes must be changed too.

Do you remember the case of Songdo City in South Korea, where technology has been broadly used to improve the *capital product* (the creation of a smart city), but failed to do so? In this case, technology did not bring any value at any level; the final product, a mere demonstration of technology, has been deemed unusable and unlivable, and so, no value has ultimately ensued from it, neither for construction companies nor for the citizens of the city. In opposition to South Korea's experience, *capital products* are now being offered on an *SBM* basis with noticeable improvements in value for end-users (Baden-Fuller *et al.* 2017). It is in line with our envisioned BM, centering on managerial rather than solely technological perfections:

Our envisioned BM follows a two-sided dyadic, solution-based arrangement – encompassing three key actors: the general contractor (the construction company, a middle agent, a conductor) and two different customer segments, up- and down-stream the construction value chain – the project owner (first customer, represented by the general contractor) and the end-user (final customer).

Before we embark upon *business model innovation*, we reserve the following few sections to clarify the nuances between the two notions of *business strategy* and *business model*, and extend our investigation to cover *business tactics* too. On the surface, these notions seem to be identical – yet, in fact, they are not, and are different constructs (Casadesus-Masanell and Ricart 2010). As put by the authors, *a business model* is a reflection of a company's *realized strategy*; "it is the direct result of strategy, but it is not strategy itself". By the same token, Magretta (2002) acknowledged that a *business model* is not the same thing as a *strategy*, despite the fact that many people use both terms interchangeably today. A *business model* describes, as a system or platform, how the pieces of a given business fit together, but it does not factor in one critical dimension of performance (e.g. competition). In reality, every company is expected to run into rivals at one point in time and, when it does, it will be the job of the *business strategy* to deal with it[8]. And so, we assert that by clearly separating the realms of *strategy*, *business models* and *tactics*, we would have access to an integrative framework based on which better *business models* could be built and further progress in the field of *business model innovation* could be made.

---

8 Ibid.

	Customer Use-Value	*Example*	*Customer Engagement Mechanism*	*Author/s*	*General Contractor Business Model*	*Impact of Technology on Business Model*
DYADIC ARRANGEMENTS (COMPRISING TWO PLAYERS)						
PBM	*Customer obtains use-value via consumption after purchase*	• A washing machine used at home • Frozen food that is cooked and eaten at home	***Transactional***: Provider puts product or service on market via a value chain with limited interaction post purchase	Vargo and Lusch (2004)	Old construction value chain	*Insignificant*
SBM	*Client obtains use-value in presence of and with help from the provider*	• A meal that is consumed in a full-service restaurant • A strategy consultant who works with the customer to co-create a solution	***Relational***: Producer and customer co-create in real time	Gronroos and Voima (2013)	New construction value chain (– based on the involvement of a *central operator*)	*Insignificant*
TRIADIC ARRANGEMENTS (COMPRISING THREE OR MORE PLAYERS)						
MMBM	*Customer obtains use-value from reductions in search effort in finding a product or service*	Renting a previously unknown apartment via an independent platform provider	An auction house that connects potential clients with suppliers, thus saving search effort	Marshall (1890), found in Komleh (2018)	***Special case I***: Whenever a real estate brokerage company is involved in the process	*Significant*
MSBM	*Customer obtains use-value from consuming a product or service and gains extra value on account of dealings with others*	• A free magazine offering content and advertisements, both of which are valued by readers • A web-based free game that contains advertising valued by the players	A platform that supplies a product or service that contains extra benefits from 3rd parties – that typically pay for the whole package	Rochet and Tirole (2006)	***Special case II***: Whenever a specialized magazine offering the option of free-of-charge advertisement is involved in the process	*Significant*

Table 3.5. *Fundamental business model arrangements (source: adapted from Baden-Fuller* et al. *(2017))*

3.4. An amalgam of strategies and tactics

A BM, we recap, is a simplified representation of how a company makes money (Casadesus-Masanell and Tarzijan 2012). Referring to the *decision-making model* shown in Figure 3.7 below, it seems quite evident that it is the fundamental set of decisions that form the business and allow for its continued existence and profitability (Lee and Stinson 2014).

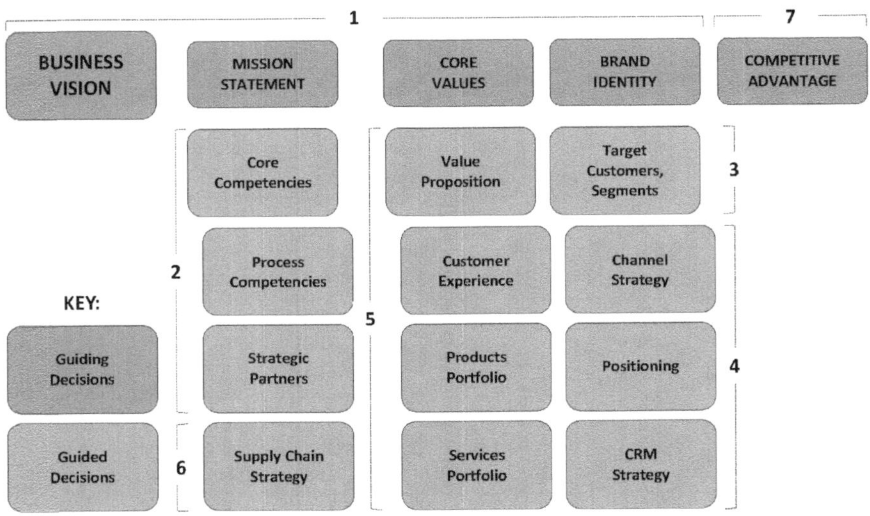

Figure 3.7. *The decision-making model (source: created by the author – inspired by Lee and Stinson (2014)). For a color version of this figure, see www.iste.co.uk/karam/general.zip*

Essentially, a company's business plan documents these key decisions, adding information that supports the alternatives chosen and demonstrating a viable BM. After viability, the company should identify how to sustain its operations in the face of rivalry (Massa and Tucci 2014).

As per Magretta (2002), choosing a competitive-advantage strategy typically identifies the areas where the company should focus its innovation (or investment) efforts to boost customer value proposition over time. To incorporate the concepts of *strategy*, *BM* and *tactics* however, we set off by introducing the *generic two-stage competitive process framework* (Figure 3.8). The rationale behind the latter, we add, is quite straightforward; it is as follows: normally, companies start by choosing a logic of (strategy for) value creation and value capture, which would be reflected in their BMs. Later, they make tactical choices driven by their intents to maximize stakeholder value. Therefore, sequentially, *strategy* comes first and *tactic* follows; and the *BM* falls in between the two. With that said, we use the term *stractics* in the next sections to denote *BMs*, being concurrently a reflection of a company's sets of strategies and an influencer of its sets of tactics.

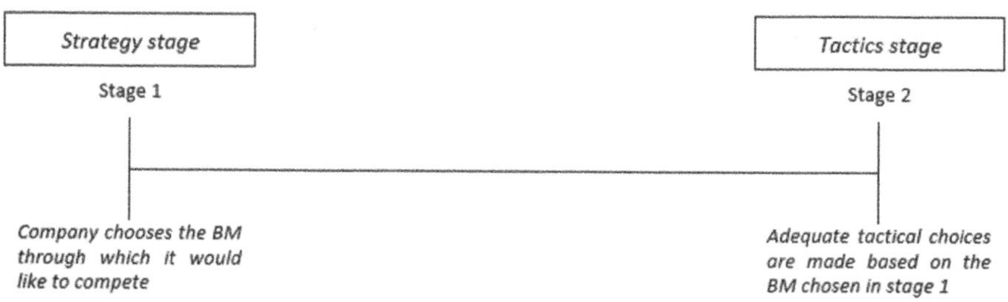

Figure 3.8. *The generic two-stage competitive process framework (source: adapted from Casadesus-Masanell and Ricart (2010))*

3.4.1. *Strategies*

A *strategy* is defined as a provisional plan of action intended to realize a precise objective. It is the art of creating value (Normann and Ramirez 1993). Moreover, as described by Eisenmann (2002), a critical element of strategy is the set of committed choices made by management. Similarly, Porter (2001) points out: "strategy is the creation of a unique and valuable position, involving different sets of activity systems".

In line with the aforementioned, in this book *strategy* refers to the provisional plan for which BM to use and at what point to use it. The word *provisional* is of great significance here, indicating that strategies include provision for eventualities that may not even arise. For example, a given company (*Y*) facing a potential entrant (*X*) in the industry would normally have as a strategy, a plan (*A*) as to what to do if (*X*) enters the market; a plan (*B*) as to what to do if (*X*) refrains from doing so. Assuredly, only one of the two options would actually take place, thus forcing (*Y*) to take necessary action. However, in the case of an *outsider* looking at (*Y*), as reported by Chesbrough and Rosenbloom (2002) and by Johnson *et al.* (2008), they would only be able to perceive the company's realized strategy (*A* or *B*), not the complete one (*A* and *B*).

Returning to the case of Ryanair, when the company was on the brink of bankruptcy back in 1991, its strategy was an action plan to transform its standard full service airline BM. Rivkin (2000), in his article "Dogfight over Europe: Ryanair (A)" richly describes the four alternative action plans that Ryanair's executives opted for to escape the imminent collapse of this business: (1) become *the* airline operator of the southwest of Europe; (2) add a business class; (3) become a *feeder* airline operating from Shannon airport; or (4) exit the industry. Aside from the fourth option, we believe all of the others demanded a different BM, and the specific way Ryanair executed these plans was through *one realized strategy* linked to *one specific BM*.

3.4.2. *Tactics*

Tactics (in comparison to *strategies*) describe the remaining choices available to a company by virtue of the *strategy*, and so, the *BM* currently in use. Consider the case of *Metro*, for instance, the world's largest newspaper in circulation (Khanna *et al.* 2007; Casadesus-Masanell and Ricart 2010). It is *free* and *ad-sponsored* (advertising-sponsored) and is published in over one hundred cities. In each city it enters, it competes with local newspapers sold at positive prices. As the *Metro* is *ad-sponsored*, it specifies the rates it charges to companies wishing to advertise in it. *Metro* also chooses the exact number of pages that each edition of the newspaper has, the exact number of ads, the precise balance between news and opinion pieces and so on and so forth. All of these choices are part of *Metro*'s *tactics*. However, *Metro* cannot amend the price of its product because its BM is *ad-sponsored*, and therefore, the product must be given away for free. In other words, *Metro's BM* prevents the company from using *newspaper price* as a parameter that could be altered depending on level of competition and other external factors. For that reason, we conclude, the latter is not part of *Metro*'s set of *tactics*.

Now, we proceed by enquiring: *Why are tactics essential anyhow*? They are because they play a crucial role in determining how much value is created and captured by the company (Casadesus-Masanell and Ricart 2010). In the case of *Metro*, advertising rates and the number of ads displayed in the newspaper ended up influencing readership-and-advertising revenues. In fact, as progressively more ads were included, readers became increasingly annoyed and less disposed to reading the newspaper. Equally, as the advertising rates rose, the count of advertisers wishing to advertise in the newspaper plunged, thus crushing *Metro*'s revenue, profit and value capture:

To further illustrate the essentiality of tactics in the business world, we raise the case of cars – where the car itself designates the BM. Certainly, the way the car is built places constraints on its use-value that is, what the driver could or could not do with it. A large, powerful 4×4 car, for instance, makes it stiff for a driver to maneuver on the narrow streets of the city of Paris. In contrast, a small, powerless compact car would create more value for the driver by making this task far less cumbersome. Thus, under specific settings, there are tactics that are possible with the compact car that would be impossible with the 4×4. Put differently, we conclude that the shape of the car – being an element of the BM – is decisive, in some particular cases at least, of its use-value and so, its overall worth.

So far, we have argued that *tactical choices* define how much value is created and captured by the company using them – however, as per Casadesus-Masanell and Ricart (2010), there is still much more to *tactics* than this – with their influence, the authors alleged, reaching out, now and then, to other companies with which the focal company interacts. At this level, we are referring to the idea of *tactical interactions*, which denote the various ways companies influence each other by acting within the bounds set by their respective BMs. Generally, *tactical interactions* arise when a company's BM is put in contact with that of another company (e.g. mom and pop stores and discount retailers)[9]. And when this materializes, there would be *consequences* on both companies' BMs – where feedback to the rest of the system is dogged not only by the focal company's *tactical choices*, but also by those made by the other company (McGrath 2010).

Finally, we affirm, a BM employed by a company determines the set of tactics available for it to compete against, or to cooperate with, other companies in the marketplace. Therefore, BMs and tactics are closely related and mutually dependent, with the likelihood of one influencing the other (Yip 2004).

3.4.3. *Stractics*

Having introduced the notions of *strategies* and *tactics* (and earlier, *BMs*), we – once again – use the *generic two-stage competitive process framework* to conceptually integrate and relate them (see Figure 3.9). As described by Casadesus-Masanell and Ricart (2010), BMs are presumed to lie in between *strategies* and *tactics*. And the linkage between the two concepts is called *stractics*,

9 Ibid.

a real-world terminology that we use in our text to designate the amalgam of *strategies* and *tactics* at the corporate level.

Precisely so, it is a company's choice of a strategy that would depict which BM to espouse – and so, which set of tactics to apply based on circumstances and needs. As formerly noted, a *strategy* is a provisional plan of action as to which BM to use, and the available actions for *strategy* are choices that constitute the raw material of BMs. Based on the *strategy*, companies define long-term goals and ways to fulfill them. Consequently, *strategy* entails designing (redesigning) BMs to allow for companies to reach their prefixed durable business objectives. With regard to *BMs* in particular, they are nothing but a plain reflection of a company's realized strategy. As for *tactics*, much like *strategies*, they are concrete action plans, courses of action that occur within the bounds drawn by the company's BM. Nonetheless, in contrast to *strategies, tactics* are generally oriented towards smaller steps and shorter timeframes.

Within this framework, and to better cement the prevalent dependencies between all three notions, we recall the *car* example put forward earlier – but this time, we suppose that the driver has been given the chance to bring all the changes he wants to the *car* (e.g. color, fuel consumption capacity, power engine and others) prior to operating it. Business-wise, the changes made (one or more, minor or major) symbolize *strategies* because they *ipso facto* alter the car (*BM*) in its entirety. And the company's *competitive process framework* would discharge in the following manner – in order:

– (1) the design and building of the car are the *strategy*;

– (2) the car itself is a *BM*; and

– (3) the driving of the car is the *tactics*.

This reasoning relates to our research project in so many ways. Mainly, by the fact that what we are proposing is an industry-level BM change, an organizational innovation (a new business strategy) that construction companies may opt for to better manage smart developments. Also, given that construction companies may have one or several BMs (moneymaking logics), we, through this study, offer them one more, plainly, a new set of *stractics* they could rely on for the resourceful achievement of their smart construction projects.

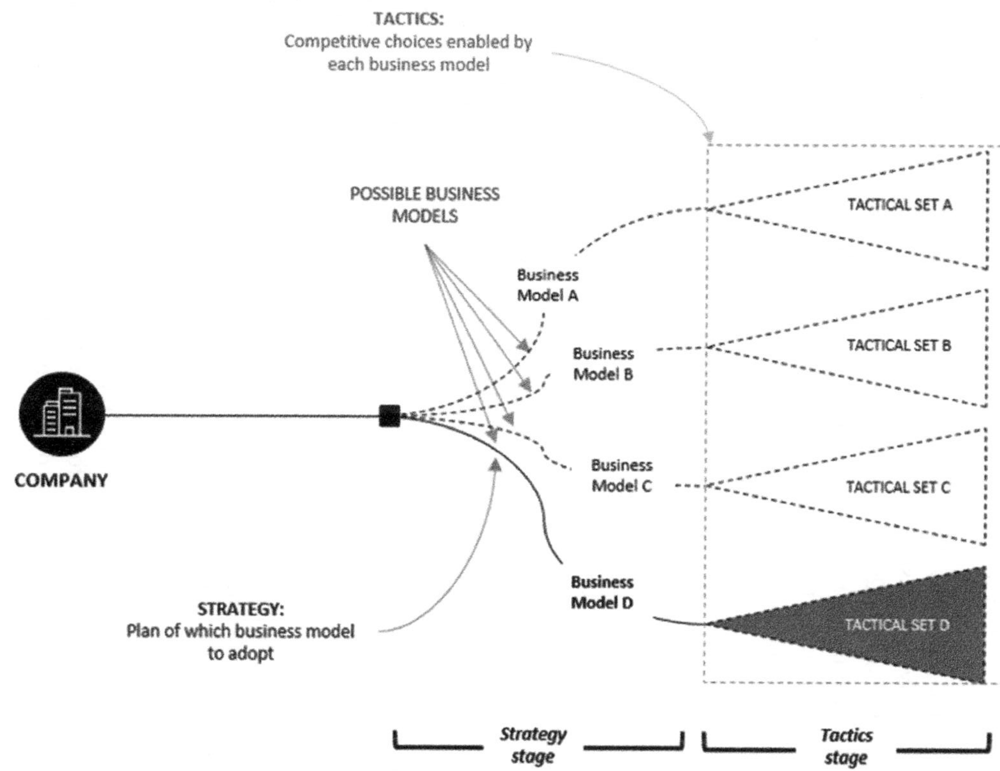

Figure 3.9. *Strategy, business model and tactics (source: adapted from Casadesus-Masanell and Ricart (2010)). For a color versionof this figure, see www.iste.co.uk/karam/general.zip*

The upcoming sections present a lengthy primer on the passage from BM to BMI (Business Model Innovation), describe the four main streams of research on BMI and underscore the main gaps in the research on BMI today.

3.5. On business model innovation: a long primer

Over the years, the scope of creating CVP (customer value propositions) has extended from delivering basic products and services to crafting advanced BMs (Stonehouse and Snowdon 2007; Massa and Tucci 2014, Laszczuk *et al.* 2017). Porter (2001) netted this paradigmatic shift by declaring that a strategic fit among many activities is crucial, not only for a competitive advantage, but also for the sustainability of that advantage. Indeed, it would be stiffer, the author stated, for a competing company to match a collection of interlocked activities than to copy a specific sales force approach, match a technology process or duplicate a set of product features. And positions built on systems of activities, the author later on concluded, are far more sustainable than those built on individual activities.

Within this context, the old dictum of *building a better mousetrap*[10] – as a means to make the world beat an innovation path to a company's doorstep – springs to mind. Knowing that all companies (evidently!) have some sort of a BM, we, as a result, assume that the key questions that should be asked at this level are the following (Morris *et al.* 2006): *Have the elements of the BM purposely been thought through following a customer-centric approach? Does a company decisively design all elements of the BM or does it focus most of its innovation on upgrading the core product?* As put by Carassus (2004), all eggs are often put into one basket, that of product innovation, with the majority of companies constantly searching for differentiation through superior products while at the same time neglecting opportunities in other parts of the BM. Actually, to grasp the entire activity system is a complex task that frequently ensues in both overlooked areas and designs that are built on the dominant logic of the industry (Massa *et al.* 2017).

We carry on by asking: *How could a company guarantee that its newly designed BM would end up delivering as planned?* According to Christensen *et al.* (2006), four areas should be investigated for a company to establish a systematic overview of opportunities, exploit BM design and enhance customer value creation (Figure 3.10).

10 After Barker, G. and Bissell, P. (2007). *A Better Mousetrap: The Business of Innovation.* abettermousetrap.co.uk. Available from AbeBooks.

Customers	*New*	Expand the market	Create new markets
	Existing	Adjust fit to existing market	Disrupt the existing market
		Incremental	*Radical*
		Business Model Change	

Figure 3.10. *Typology of business model innovation opportunities (source: created by the author – inspired by Christensen* et al. *(2006))*

It is all about adjusting the current BM to exploit opportunities in existing markets. In fact, the shifting needs of existing customers and a turbulent environmental context constitute a window of opportunity for companies to amend their BMs (Chesbrough 2010). Under similar settings, simple incremental moves in some areas of the BM could be subjugated to achieve significant business goals. *TDC* (Tele Denmark Communications)[11] is a great example of a company which, via the introduction of a free-music service layer, has vividly changed the stickiness of its BM, and diminished loss in market shares via improved customer loyalty and satisfaction. And so, as informed by Souto (2015) and Foss and Saebi (2017), changes brought to the table for the creation and design of new BMs do not necessarily always have to be radical – because simple moves, time and again, suffice to clutch new differentiation factors and accomplish great results in existing markets. But then, *what about redesigning the current BM to expand the market?*

Incremental change in the current BM may be leveraged to expand the market by attracting non-customers, thus converting latent demand into real demand (Lindgart *et al.* 2009; Koen *et al.* 2011). The practice of serving products in small, reasonable sizes is a broadly used strategy by global consumer brands to enter emerging markets. In the Philippines, for instance, 7 out of 10 smokers purchase their cigarettes by the stick rather than by the pack; also, 8 out of 10 Filipinos buy their shampoos by sachet rather than bottles (Rodolfo and Sy-Changco 2015). Here again, similarly to *TDC*'s case, the core product is left intact, while both innovative distribution setups and packaging designs became recipes for business success. *So, now, how about tossing in a new BM that would disrupt the existing market?* In some cases, inspecting the needs of current customers might lead to the discovery

11 European Commission and DG Research & Innovation (2013). The Need for Innovations in Business Models. Final Policy Brief (Deliverable 5) [Online]. Available at: http://ri-policy-analysis.eu/fileadmin/ripolicies/documents/bmi/policy_brief/eriab-bmi_pb_new_business_models.pdf [Accessed May 27, 2021].

of attractive sub-segments which are not suitably served by the current BM (Kumaraswamy *et al.* 2018). Certainly, the detection of underserved or overserved customer segments could pave the way for important managerial decisions. This is chiefly the case of Nestlé's Nespresso[12] which, by eliminating the risk of failing to brew the perfect cup of coffee, and decreasing the time it takes to make a nice single-shot espresso, succeeded to identify and acquire an attractive niche of espresso drinkers (Söderberg 2019). In contrast to high-end disruptions typically achieved via notable technology leaps, low-end disruptions, on the contrary, are accomplished via simple cost-cutting measures (e.g. Ryanair).

Moving forward, we enquire: *What about designing entirely new BMs?* Schiavi and Behr (2018) precised that designing new BMs could be leveraged to create wholly new markets where large groups of customers have been, intentionally or unintentionally, locked out. Nevertheless, the authors added, in some specific cases, minor BM tweaks such as sachet marketing are just not enough. And so, a complete overhaul is required at times for a company to commercialize CVP for new markets. Table 3.6 shown below describes the growth opportunities that could be grasped through BMI. These opportunities, we say, could be explored and exploited by any company to sustain growth and delight customers. As per Markides (2013), introducing a better BM into an existing market is the definition of a disruptive innovation. Obviously, new BMs are not launched out of the blue, the author indicated, but rather closely aligned with overarching the strategic growth ambitions of the company. And so, the author added, a diagnosis of the strengths and weaknesses of current BMs must form the basis for any thoughtful attempt to pursue any generic growth opportunity.

To date, several scholars have suggested signs that could indicate that a company's current BM is running out of gas. According to McGrath (2010) for instance, the first symptom is when innovation to a company's current offerings create smaller and smaller improvements. In such as case, the author suggested, BMI is the only solution available. Similarly, Foss and Saebi (2017) affirmed that BMI is a crucial route for companies seeking to reinvigorate a lagging core, defend against industry disruption or decline, or even drive breakthrough growth. Amazon[13] is a good example of a company that drove BM change in a resourceful way. At the start, the company was the Earth's biggest bookstore; 20 years later, that bookstore turned into a leader in cloud computing, and started delivering groceries and producing Emmy Award-winning television series.

---

12 Ibid.

13 Amazon.com (2018). Harvard Business Review Case Collection [Online]. Available at: https://www.hbs.edu/faculty/Pages/item.aspx?num=49324 [Accessed May 27, 2021].

	ADJUST FIT TO EXISTING MARKET	DISRUPT THE EXISTING MARKET	EXPAND THE MARKET	CREATE NEW MARKETS
Customers	Existing	Existing	New	New
Level of change	Incremental	Radical	Incremental	Radical
Characteristics	Improve the existing BM to attract and retain customers in existing markers characterized by intense competition	Develop new BM to exploit opportunities in either underserviced high-end segments or overserviced low-end segments	Improve the existing BM to fit needs of customers in unserved markets or segments of non-customers	Develop new BM to exploit opportunities in new markets or segments of non-customers

Table 3.6. *Generic growth opportunities (source: created by the author, inspired by Markides (2006, 2013))*

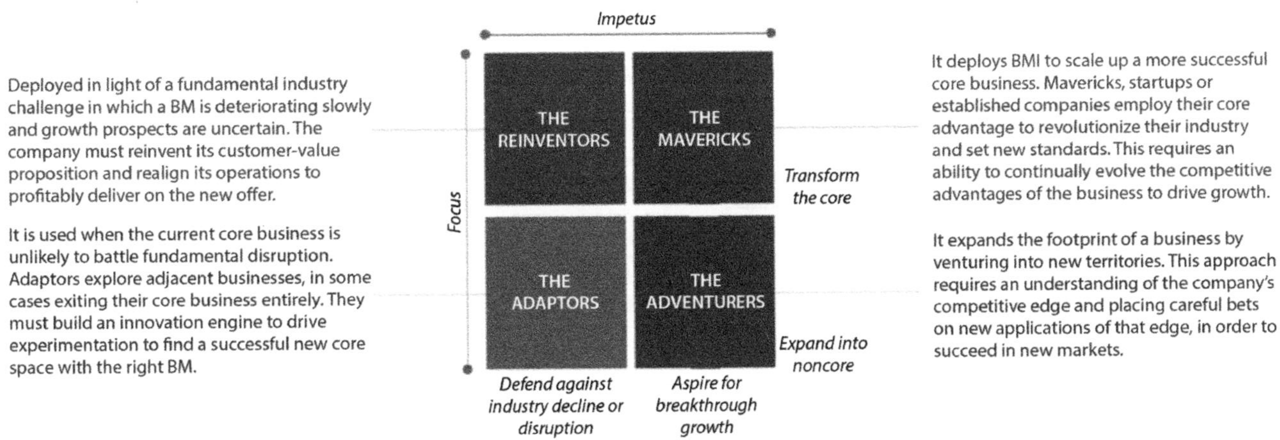

Figure 3.11. *The four approaches to business model innovation (source: inspired and adapted from Deimler and Kachaner (2020)). For a color version of this figure, see www.iste.co.uk/karam/general.zip*

Unlike other types of innovation, changes to the BM infer changes to the foundational decisions upon which the business operates (Massa and Tucci 2014) – meaning that BMI is most likely to be radical and transformational. Changing the entire BM design (relative to incremental innovation: mere product innovation or technology enhancements) is possibly to be associated with higher risks, the authors said, with the possibility of disrupting the current business. Hence, "*to find the right timing for a BM change is fundamental to business success*", they concluded. For well-established businesses, recognizing and managing this kind of radical change could be critical to their long-term survival, whereas for start-ups, such a change may be advantageous. With that said, although present-day changes in customer behavior, globalization and technological innovation have left the door wide open for the advent of new BMs (Schiavi and Behr 2018), we must realize that not all BMI efforts are alike (see Figure 3.11 for an illustration of the various approaches to BMI).

3.5.1. *Definitions*

To date, the BM concept has been embarked upon in the fields of *innovation* and *technology management*, with two chief ideas typifying the academic research pertaining to it: (1) companies promote original ideas and technologies via their BMs; and (2) the BM is a new substance of innovation by itself. Chesbrough and Rosenbloom (2002) advanced the case of *Xerox* to pinpoint the ability of a BM to unlock the value potential embedded in new technologies and convert it into a concrete outcome. From their side, Calia *et al.* (2007) affirmed that BMs could not only entail consequences for technological innovations, but could also be shaped by them. Moreover, as put by Johnson and Suskewicz (2009), BMs play a vital role in the commercialization of technologies at two levels, that of an individual company, and of an entire industry. In large, in infrastructural change, the authors argued, the key intent is to shift the focus from developing distinct technologies to creating completely new systems. The BM in such a case is introduced as part of an all-inclusive framework for thinking about systemic change. Although the BM looks like a terrific thing, we recap that studies on BM, innovation and technology management have shown that technology might not be enough by itself to guarantee business success (Doganova and Eyquem-Renault 2009), for it lacks inherent value (Chesbrough 2007).

Apart from embedding technology in outstanding products or services, Mitchell and Coles (2003) said that a company should design a unique BM to fully grasp its operating potential. Indeed, as per the authors, companies will possibly view the BM itself as a substance of innovation – hence, the notion of *open innovation* introduced by Chesbrough (2003). Generally, *open innovation* – being a mode of innovation in which companies look outside their boundaries in order to leverage internal and external sources of ideas – entails the espousal of new BMs designed for sharing or licensing technologies (Chesbrough 2010). Today, there is a swelling agreement that

BMI is decisive of business performance (Foss and Saebi 2017), and the number of scholars investigating BMI as a means for corporate conversion and rejuvenation is on the rise (Johnson *et al.* 2008; Demil and Lecocq 2010).

At this point, now that we have concisely presented the concept of BMI, it seems opportune to define it. Overall, we proclaim, the definitions that we could come across while reviewing the academic literature on BMI diverge greatly, thus revealing some deep vagueness with regard to what the latter really is (Foss and Saebi 2017).

For some scholars, BMI is a *process* – while for some others, it is more of an *outcome*. Such variances, we proclaim, are expected to have significant implications for subsequent research on BMI. Indeed, studies that perceive BMI as a *process* often take a dynamic approach and look into the managerial physiognomies that facilitate or hinder the process of BMI (Demil and Lecocq 2010; Doz and Kosonen 2010), whereas studies focusing on the *outcome* tend to be more descriptive and identify the content of the BMI *ex post* (Bucherer *et al.* 2012; Günzel and Holm 2013). As put by Foss and Saebi (2017), even though both types of studies have their merits, they do deal with distinct phenomena, each requiring a different empirical approach to be scrutinized. Like for BMs, the definitions suggested for BMI are numerous (see Table 3.7), yet they tend to lack specificity. This, again, infers that there is no clarity in the literature about the nature of a BMI.

Definitions	***Author/s***
BMI means BM replacements that provide product or service offerings to customers and end-users that were not previously available. The process of developing these novel replacements is also referred to as BMI.	Mitchel and Coles (2004)
BMI is the discovery of a fundamentally different BM in an existing business.	Markides (2006)
BMI is a reconfiguration of activities in the existing BM of a company that is new to the product service market in which it competes.	Santos *et al.* (2009)
Initiatives to create novel value by challenging existing industry specific BMs, roles and relations in certain geographical market areas.	Aspara *et al.* (2010)
BMI occurs when a company adopts a novel approach to commercializing its underlying assets.	Gambardella and McGahan (2010)

Definitions	*Author/s*
BMI is about generating new sources of profit by finding novel value proposition/value constellation combinations.	Yunus *et al.* (2010)
Innovate BMs by redefining: content (adding new activities), structure (linking activities differently), and governance (changing parties that do the activities).	Amit and Zott (2012)
BMI is the process that deliberately changes the core elements of a company and its business logic.	Bucherer *et al.* (2012)
A BMI happens when the company modifies or improves at least one of the value dimensions.	Abdelkafi *et al.* (2013)
Corporate BM transformation is defined as a change in the perceived logic of how value is created by the company, when it comes to the value-creating links among the company's portfolio of businesses, from one point of time to another.	Aspara *et al.* (2013)
A BMI could be thought of as the introduction of a new BM aimed to create commercial value.	Berglund and Sandström (2013)
BMI refers to the search for new logics of the company and new ways to create and capture value for its stakeholders; it focuses primarily on finding new ways to generate revenues and define value propositions for customers, suppliers and partners.	Casadesus-Masanell and Zhu (2013)
BMI refers to (a) the design of novel BMs for newly formed companies (i.e. the activity of creating, implementing and validating a BM), or (b) the reconfiguration of existing BMs for established companies (the phenomenon by which executives reconfigure organizational resources and acquires new ones). Both phenomena are change phenomena and could lead to BMI.	Massa and Tucci (2014)
BMI activities could range from incremental changes in individual components of BMs, extension of the existing BM, introduction of parallel BMs, right through to disruption of the BM, which may potentially entail replacing the existing model with a fundamentally different one.	Khanagha *et al.* (2014)

Table 3.7. *Definitions of BMI (source: adapted from Foss and Saebi (2017))*

For the purpose of our research, we adopt the definition suggested by Pavie *et al.* (2013) for BMI. According to the authors, BMI is the art of enhancing advantage and value creation by making simultaneous, mutually supportive changes both to a company's CVP, as well as to its underlying operating model. *At the value proposition level*, the authors affirmed, these changes could address the choice of the target segment, product and service offering, and revenue model. While, *at the operating model level*, they added, the focus is more on how to drive profitability, competitive advantage and value creation. Put differently, BMI describes a fundamental change (technology-driven or not) in how a company delivers value to its customers.

The next section exposes the four main streams of research on BMI.

3.5.2. *Streams of research on BMI*

Although there is today a noteworthy number of studies that revise the existing literature on BMI (Zott *et al.* 2011; Lambert and Davidson 2013; Wirtz *et al.* 2015; Foss and Saebi 2017), there is only one that we believe to be complete and exhaustive – that written by Schneider and Spieth (2013), which appraises over 30 scientific papers on BMI. Foss and Saebi (2017), too, did a great job gathering and summarizing data linking to BMI research. Specifically, they reviewed nearly 150 publications in terms of their conceptual, theoretical, and empirical developments and contributions (Table 3.8), and distinguished four fairly overlapping streams of BMI research (Table 3.9). Those streams, the authors added, embody weighty evolutions, as well as limitations in the arena of BMI research in general. They are discussed as follows.

The first stream, *conceptualizing BMI*, centers on the phenomenon of BMI itself, suggesting definitions and conceptualizations of BMI (e.g. Teece 2010; Amit and Zott 2012). It also focuses on issues such as the least significant definition of BMI, and the dimensions along which companies could innovate their BMs (e.g. Santos *et al.* 2009). At this level, the purpose seems to revolve around the development of definitive, pictorial schemes; nevertheless, as formerly noted, most definitions available, though abundant, vary strikingly and are often unclear. When it comes to the second stream, *BMI as an organizational change process*, innovation is said to strongly defy prevalent managerial structures (Damanpour 1996). In contrast to the first stream, the second one goes farther by underscoring the competences, governance and scholastic systems required for prolific BMI. Indeed, studies carried out within the second stream define BMI as a vigorous process by: (1) recognizing the various stages of the BMI process (e.g. Gassman *et al.* 2014); (2) categorizing the different administrative know-hows and procedures needed to support such a change process (e.g. Demil and Lecocq 2010); (3) revealing the significance of research and knowledge (e.g. Andries and Debackere 2013); and (4) putting practical tools forward for the implementation of the process (e.g. Evans and Johnson 2013).

Streams	*I*	*II*		*III*	*IV*
Focus	***Conceptualization of BMI:*** *case examples*	***BMI as an organizational process:*** *case examples*	*Case studies:*	***BMI as an outcome:*** *case studies*	***BMI and managerial consequences:*** *survey data*
Author/s	• Markides (2006) • Johnson et al. (2008) • Santos et al. (2009) • Koen et al. (2011) • Sorescu et al. (2011) • Amit & Zott (2012)	• de Reuver et al. (2009) • Deshler & Smith (2011) • Berglund & Sandstroem (2013) • Evans & Johnson (2013) • Cavalcante (2014) • Girotra & Netessine (2014)	• Demil & Lecocq (2010) • Deshler & Smith (2011) • Pynnonen et al. (2012) • Enkel & Mezger (2013) • Aspara et al. (2013) • Frankenberger et al. (2013) • Dmitriev et al. (2014) • Khanaga et al. (2014) • Mezger (2014)	• Anderson & Kupp (2008) • Gambardella & McGahan (2010) • Sánchez & Ricart (2010) • Yunus et al. (2010) • Wirtz et al. (2010) • Berman (2012) • Holm et al. (2013) • Richter (2013)	• Pohle & Chapman (2006) • Zott & Amit (2007) • Aspara et al. (2010) • Bock et al. (2012) • Huang et al. (2013) • Denicolai et al. (2014) • Wei et al. (2014) • Velu & Jacob (2014) • Kim & Min (2015)

Table 3.8. *Streams of BMI research (source: adapted from Foss and Saebi (2017)). For a color version of this table, see www.iste.co.uk/karam/general.zip*

Author/s	*George & Bock (2011)*	*Zott et al. (2011)*	*Lambert & Davidson (2013)*	*Wirtz et al. (2015)*	*Schneider & Spieth (2013)*
Findings	*Use of BMs:* • Organizational design • Resource-based view • Narrative and sense making • Nature of innovation • Transactive structure • Opportunity facilitator	*3 themes of BM literature:* • E-business • BMs and strategy • Innovation and technology management	*3 themes of BM literature:* • BM as basis for enterprise classification • BMs and enterprise performance • BMI	*4 research foci:* • Innovation • Change and evolution • Performance and controlling • Design	*3 streams of BMI research:* • Prerequisites of conducting BMI • Process and elements of BMI • Effects achieved through BMI

Table 3.9. *Research on BMs and BMI (source: ibid)*

Moving forward, the third stream, *BMI as an outcome*, embarks upon the yield of the second stream. Commonly, it addresses the advent of new BMs in a specific sector or industry (e.g. Schneider and Spieth 2013; Souto 2015). Also, some studies within this stream concentrate on scrutinizing a precise type of new BM – one that is conceived for low-income markets, for instance (e.g. Sánchez and Ricart 2010). Other publications within this stream (e.g. Matzler *et al.* 2013), we assert, tend to be descriptive rather than exploratory, limiting their contributions to commentaries on the creativity of some prominent companies' BMs (e.g. Nestlé's Nespresso). For some reason (one or more), the third stream does not build on the first one, yet, instead, the emphasis is put on unfolding one specific type of change in BMs – often said to be of a new kind. This stream is limited; it does not provide any discussion of the standards based on which the proper BM change could be seen as new.

Finally, the fourth stream, *the managerial consequences of BMI*, tackles the managerial performance repercussions of BMI. In this stream, we could distinguish between studies that link the BMI process to outcome implications (e.g. Cucculelli and Bettinelli 2015), and those that inspect the influence of various types of BMs on company performance (e.g. Zott and Amit 2008). In the first case, studies assume a process view and investigate whether an innovative change in the existing BM leads to larger performance outcomes (Giesen *et al.* 2007; Aspara *et al.* 2010; Cucculelli and Bettinelli 2015). In the second case however, studies do not directly link BMI to performance outcomes; instead, they test, following an empirical approach, the impact of various BM designs on innovation performance (Zott and Amit 2008; Wei *et al.* 2014).

To say the least, the BMI literature has so far yielded several key insights, allowing for a better understanding of the nature of BMI in general. One central contribution of BMI research links to the ability of companies to incorporate change in the design and architecture of their BMs as a strategic vehicle for the creation of both value and competitive advantage (Massa and Tucci 2014). Within the scope of our investigation, two lines of argumentation seem to be obvious; academic researches either follow a dynamic view of BMI or conceptualize it as a managerial change process (*first stream*), or view BMI as a new type of innovative schemes (*third stream*) that could ultimately influence business performance (*fourth stream*). We conclude by pointing out that the literature on BMI lacks academic works that plainly deal with the antecedents of BMI – yet, we note that the four research streams discussed earlier have largely evolved over time, perhaps disconnectedly, because no direct correlation could be spotted among them (Foss and Saebi 2017).

In connection with the present section, the next one demarcates the gaps in the academic literature on BMI.

3.5.3. *Gaps and challenges*

According to Foss and Saebi (2017), the existing literature on BMI is highly fragmented, and the gaps in the studies carried out over the years to investigate it are copious. For simplicity reasons, the identified gaps are categorized under six specific headers (themes), namely: (1) *definitions*; (2) *dimensions*; (3) *antecedents*; (4) *outcomes*; (5) *variables*; and (6) *boundaries* (see Table 3.10). They are clarified hereunder.

Definitions: up until now, we have plainly exposed the various inconsistencies in the definitions suggested for BMI, pointing altogether to a lack of clarity and specificity in the literature stream regarding the nature of BMI. Although this might not be a bad thing, a characteristic of a nascent research field that has not formed yet, such conjectural heterogeneity, we believe, may be challenging in the long run since research efforts would end up branching off in various, random directions[14].

Dimensions: advances in research frequently occur when constructs are clearly dimensionalized, that is, when scholars succeed to capture the heterogeneity of a construct in terms of its key components that have relevant inferences for outcomes. Our review of the literature suggests that BMIs differ in terms of at least two dimensions: *scope* and *novelty*. The degree of novelty of BMI relates to whether BMIs are seen as novel at the company or industry level (e.g. Johnson *et al.* 2008; Santos *et al.* 2009). As for its scope, reflecting how much a BM is reflected by BMI, it could start small (BMI influencing only one component of a BM) (e.g. Amit and Zott 2012; Schneider and Spieth 2013), yet grow larger and larger over time (BMI influencing two components, three, four, or even all components of a BM – plus the architecture connecting them) (e.g. Yunus *et al.* 2010).

Antecedents: to our knowledge, there is little or no publications that theorize the antecedents of BMI. Actually, most of the studies that we bumped into while inspecting the matter seemed to tackle the BMI concept from varied slants. Doz and Kosonen (2010), for instance, acknowledged that BMI is a compulsory retort

14 Ibid.

to strategic incoherence and disorders, convergence and strong global competition. In contrast, Johnson *et al.* (2008) stated that a company should opt for BMI whenever it is facing competitive pressure or an unstable base of rivalry. Voelpel *et al.* (2004), rather, talked about the prominence of BMI whenever chief and erratic changes in the business context have emerged. Some other studies (see Tables 3.8 and 3.9), we add, described BMI as an endeavor, a strategy often used by companies, to grasp new opportunities announced by the advent of new technologies. Also, a few others focused on assessing the impact of new BMs on business performance in the context of e-business (Pateli and Giaglis 2005; Sabatier *et al.* 2012). The list of scholars attempting, say, to identify the use-value of BMI in the corporate world stretches endlessly, yet, what is worth mentioning at this point is that most of the relevant academic works carried out on this basis have been, as put by Foss and Saebi (2017), backdated and inductive – rather than deductive and hypothetical.

Outcomes: our investigations led us to recognize that the influence of BMI on business performance is likely to vary based on the improvements or strategies that companies choose to adopt in order to innovate their BMs (e.g. cost cutting, process optimization and launching new products). Generally, there are only few articles that clearly explicate how BMI boosts profitability (e.g. Aspara *et al.* 2010; Wei *et al.* 2014). One reason for that lies in the intricacy of linking BMI and performance. This is genuine because, often, there might be numerous composite links between BMI and performance, links that play out differently across time with the possibility of being entangled. Thus, it seems impractical for scholars to empirically account in their studies for all the inherent mechanisms governing the relation between BMI and corporate performance (Foss and Saebi 2017).

Variables: while revising the literature on BMI, we have found numerous studies that point towards the crucial roles that *managerial capabilities*, *leadership skills* and *educational processes* could play in provoking BMI; also, we have recognized that some other, equally important, managerial variables were, regardless of the reason, only meagerly addressed. To start with, Demil and Lecocq (2010) stressed the importance of building *dynamic consistency*, which the authors described as the ability of a company to prompt and sustain business performance while varying its BM. Similarly, Doz and Kosonen (2010) focused on the ability of companies to achieve *strategic agility* for hastening BM renewal.

Definitions	*Dimensions*	*Antecedents*	*Outcomes*	*Variables*	*Boundaries*
• Concept • Construct	• Scope • Novelty	• BMI precursors	• Causality relationship: BMI, business performance	• Capabilities, leadership • Learning, experimentation • Cognition • Organizational structure design	• Current market standing: status quo • Adaptability: market changes, requirements
• BMI: construct of BM • BM: architecture for value creation, capture, delivery • Complex systems: functional relations, underlying activities	BMI typology: • Evolutionary BMI • Focused BMI • Adaptive BMI • Complex BMI	• Internal environment • External environment	• Business model • Governance structure • Value chain • Revenue model • Imitability, complexity	• Moderating factors	• Entrepreneurship • Open innovation • Servitization • Sustainability

Table 3.10. *Gaps in BMI research (source: created by the author – inspired by Foss and Saebi (2017))*

Going forward, other scholars (e.g. Andries and Debackere 2013) have shed light in their works on the vitality of *experimentation* and *education* as sources of BMI. Indeed, they suggested that companies could innovate their BMs via either incremental or radical experimentation modes, meaning that innovation is not always stirred by radical change, and that, in specific cases, small changes may be satisfactory to induce the required change. *Cognition*, too, seems to play an important role in instigating BMI. For Doz and Kosonen (2010), BMs stand as cognitive structures providing a scheme of how to set boundaries to the company, how to create value, and how to shape its internal structure. Besides this, other managerial factors at play in the management of both BMs and BMIs have recently also been mentioned and described in the literature (Markides 2013; Schneider and Spieth 2013), explicitly: *strategic flexibility* and *culture*. However, some other, equally important contingency variables were neglected or less tackled in the literature, the most important one, we proclaim, being *organization structure design*. Indeed, as put by Foss and Saebi (2017), the extent to which managerial design and control procedures need to be changed to support BMI, and the extent to which a BMI necessitates a new managerial design are subject matters that must be tackled by scholars in far more depth.

Boundaries: the literature on BMI does not openly contend with the issue of boundary conditions; however, these are significant, as companies may diverge with regard to the antecedents and magnitudes of BMI, based on whether they are entrepreneurial or incumbent, high-tech or traditional, start-up or established, and single industry or diversified. For instance, as previously evoked, while startups may have the suppleness required to engage in BMI, the performance repercussions of BMI may be much more evident in well-established companies (e.g. Zott and Amit 2008). Also, while research recognized that BMI is a vital vehicle for managerial makeover and renewal among incumbent companies (e.g. Demil and Lecocq 2010; Doz and Kosonen 2010), only a few studies have thus far compared BMI between incumbent and entrepreneurial companies. On the contrary, concerning sustainability issues in BMI in particular, it is only recently that these have started to gain momentum. As a result, a snowballing number of studies have suggested that sustainability objectives may call for BMI, and several scholars have started investigating the factors and/or BM designs that may possibly help companies to achieve triple-bottom-line goals: *social*, *environmental* and *financial* (Bocken 2015).

The next section, preceding the conclusion, is added to provide some real world, empirical insights into BMI. Complementing our theoretical research, it offers some straightforward examples of how BMI is (or should be) used in practice.

3.5.4. *Some practical insights into BMI*

BMI is among the most operative ways for companies to set themselves apart from the crowd – and therefore protect the existence of their businesses (Casadesus-Masanell and Ricart 2010, 2011). Specifically, it is a question of thin slicing a company into its building blocks, scrutinizing and valuing them, re-inventing them – and, in conjunction with other, new building blocks, to collate them scientifically[15]. At its meekest, BMI requires neither new tools nor the formation of new markets; it is simply about manufacturing existing products using existing technologies and delivering them to existing marketplaces through existing channels (Chesbrough 2010). And because it frequently includes changes that are imperceptible to the outside world, it could convey hard-to-copy advantages. The issue, to be clear, is to determine, with some accuracy, what BMI really entails. In the absence of a frame of reference for detecting opportunities, it would be uneasy to be methodical about the process, which expounds why it is commonly done on a case-by-case basis.

In addition, given the fact that any BM is basically an assortment of key decisions that jointly define how a business generates its profits, incurs its costs and mitigates its risks, we perceive innovations to the BM as changes to those decisions: *what* mix of products and services should be offered, *when* key decisions should be made, *who* the best decision-makers are, what they actually do, how to choose them and *why* (Betz 2002; Carassus 2004; Baden-Fuller and Morgan 2010). Resourceful changes along these dimensions, we believe, would unquestionably boost a company's blend of financial aspects. We hereunder list some practical (international) insights into BMI, that is, how innovative BMs are commonly built – and based on what rationality. These insights are crucial as some of them would directly inform the BMI that we propose at the level of the French construction industry in particular (and the construction industry globally), namely: *the engagement of a maven decision-maker* – which, in our case, is the GC (general contractor).

3.5.4.1. *What mix of products or services should be offered?*

Uncertainty, especially if linked to demand, is a serious operational impediment that all companies face, and is generally their primary source of risk. One way, among others of course, to significantly mitigate that risk is to alter the company's mix of products or services. Indeed, companies seeking to adjust their mix of offers have three options to play with and choose from. They are listed below.

---

15 Girotra, K. and Netessine, S. (2014). Four Paths to Business Model Innovation. *Harvard Business Review* [Online]. Available at: https://hbr.org/2014/07/four-paths-to-business-model-innovation [Accessed June 15, 2021].

3.5.4.1.1. Look for cohesions across offerings

The triumph of VW (Volkswagen) is chiefly due to its competent business strategy through which exist the commonalities between its various car models. Even though VW's strategy does not necessarily protect it from broad demand swipes, it cuts demand unpredictability for singular components, because shared components make it stress-free for the company, whenever needed, to shift production at its plants from one model to another. Cohesions (or similarities) are not simply shared components among different products – as they could also be, under specific scenarios, the skills required to serve several products, customers and customer segments. In the late 1990s, for instance, Amazon stretched its business from books into music, video and games – all of which necessitated the same logistics abilities that books did[16]. This has enabled Amazon to conceal the risk of failing to procure sufficient share in either of these categories with possibly larger shares in the other ones.

3.5.4.1.2. Build a hedged portfolio

Like banks, companies could choose assortment sets of products or markets to shrink the general hazardousness of their BMs. Chile's LATAM airlines, previously known as LAN airlines, opted for a somewhat similar approach[17]. Unlike the majority of US carriers (major ones included, which stem less than one-tenth of their financial returns from cargo), LATAM decided to use its planes, hovering over international routes, to transport not only passengers, but cargo too. With almost all travels from the Americas to Europe taking place on overnight flights, passenger-only airlines tend to keep their planes on the ground for long periods of time, an opportunity cost that LATAM decided to avoid by operating its planes at full capacity, whereby using the downtime to carry cargo. In fact, this approach diminished the risks related to LATAM's capacity decisions. Generally, airlines make such decisions occasionally – by ordering new airplanes – a request that is hardly achievable, thus leaving them exposed to phases of over or underutilized capacity, with some negative repercussions on revenue. And so, to hedge passengers with cargo could alleviate this risk since their respective demand curves hardly ever rise or fall together. Furthermore, to carry cargo could allow for the airline to fly cost-effectively with less passengers on board and to serve terminuses that other airlines have a tendency to avoid.

---

16 Amazon.com, 2018, Harvard Business Review Case Collection, August. Available at: https://www.uwp.edu/learn/departments/business/upload/MBA-PLLG-6-Amazon-Case.pdf [Accessed March 13, 2020].

17 LATAM Business Strategy. Available at: http://memoria2012.marketinglan.com/ingles/estrategia_y_negocios.html [Accessed March 13, 2020].

3.5.4.2. *When should key decisions be made?*

Frequently, decisions should be made *ex ante* a given company's sufficiently compiled data to make them with certainty. There are three strategies, we would say, that, based on prevailing settings, could mend a BM by shifting the timing of decisions:

– *Delay decision-making*

In lots of industries, companies tend to decide on pricing even before they start selling any of their products and/or services. This, surely, repeatedly exposes them to high risks. For instance, it seems dicey to price airplane seats prematurely simply because demand in this case is likely to vary based on prevailing economic and other conditions – and so, would unquestionably change over time.

Back in 1980, American Airlines[18] deciphered this issue by using the booking system, *SABRE* (Semi-Automated Business Research Environment), which makes it fairly easy to adjust prices swiftly by considering up-to-the minute data. Indeed, the ability of airlines to price dynamically, in real-time, has transformed the industry forever. Just recently, Uber[19], an e-business that pairs customers who need car rides with automobiles for hire, loaned out their business logic, which is in high-demand days. The company gears surge pricing, whereby prices for rides increase, thus plummeting demand while boosting supply.

– *Change the order of decisions*

Not all companies have the privilege to alter the timeline within which they operate; instead, they could jostle the order in which decisions are usually made to adjourn investment obligations until relevant information is identified. Most product development, for instance, starts with suggesting a technology (solution) for a customer need. If, *ex post* primary investments, the technology shows to be unsuccessful, then it is brought back to the drawing board (a vicious, feeble operative circle!). Yet some ICT companies, InnoCentive and Hypios for example, have recognized that if the sequence is switched to performance first and investment second, the risk associated with R&D could be as a result pointedly mitigated (Marjavonic *et al.* 2012).

---

18 Business Model Navigator (2017). American Airlines [Online]. Available at: https://businessmodelnavigator.com/case-firm?id=9 [Accessed June 15, 2021].

19 Jungleworks (2017). How Uber Works: Insights into the Business & Revenue Model [Online]. Available at: https://jungleworks.com/uber-business-model-revenue-insights/ [Accessed June 15, 2021].

A comparable variation in sequence enlightens the triumph of LiveOps[20], a company in the call center industry. Typically, conventional call centers make forthright investments in facilities and hard set-up before making any call and *ex ante* the signature of a single contract. They also choose the number of agents to be hired, the levels of talent and expertise needed, and to whom training should be provided. Afterwards, they sign up clients whose requirements match the competences they gathered. Finally, they come up with daily and weekly recruitment schemes to guarantee that sufficient operators with the fitting talents would be open to take calls. LiveOps, on the contrary, hires operators as the calls come in. Its operators work freely from home, and signal LiveOps whenever they are ready to take calls. Their pay is determined based on their performance (skill at meeting callers' needs), as well as on the duration of the call (this is made possible because all calls are recorded and scored by LiveOps). Subsequently, the company's smart software transfers callers to the most competent operators available, conferring to the nature of the call, so that capability and recruitment are regularly attuned to meet actual needs.

– *Thin-slice key decisions*

It has been a few years now since the corporate innovation and start-up worlds were taken by storm by a slightly new movement, *the lean start-up movement*[21]. At the core of the movement is a novel tactic for executives who have to make decisions about their businesses. Formerly, instigating a risky new venture entailed assembling a comprehensive business plan that covers all vital elements of the BM, and then performing according to said plan. Basically, all key decisions used to be made promptly and candidly, all at once, prior to the initiation of business activities. In contrast, the lean start-up method consists of slicing up key decisions, as follows: a venture commences with comparatively vague and imperfect assumptions about where an opening may be found; numerous stages of data collection and information gathering follow, as the BM is brushed up to reach its final, authenticated form. Characteristically, the concerned executives have a tendency to drastically modify their initial assumptions as the venture develops and new business ideas are brought to the table.

20 INSEAD (2014). LiveOps: The Contact Centre Reinvented [Online]. Available at: https://publishing.insead.edu/case/liveops [Accessed June 15, 2021].

21 Harvard Business Review (2013). Why the Lean Start-Up Changes Everything [Online]. Available at: https://hbr.org/2013/05/why-the-lean-start-up-changes-everything [Accessed June 15, 2021].

3.5.4.3. *Who are the best decision-makers?*

It is said that companies believe that decision-making in the value chain could be profoundly enriched if the people who usually make the final calls are changed. Pragmatically, concerned companies could do the following:

– *Engage a maven decision-maker*

The entire idea revolves around the idea of endowing the right, most informed persons (one or more) with the right and power to make the calls. Google's engineers, for instance, are given the liberty to choose which development projects the company should undertake, because the company trusts they are better versed in technologies and current market needs than their counterparts, the first-rank executives, are.

We note that maven people are not necessarily part of the concerned company's own workforce. Walmart, for example, about 30 years ago, decided to reassign some decision rights about stocking its store shelves to P&G, believing that the latter had, at the time, the right mixture of data and incentives to optimize its delivery and production schedules, as well as be able to keep its shelves fittingly stocked around the clock. This has indeed become a trend that we can see more in some industries relative to others. Moreover, with technological advances, some companies have been delegating their decision-making processes to algorithms. Such trends can be seen in the restaurant and catering business.

Raising the case of the restaurant chain *Not Your Average Joe's*, the latter uses *Muse* for the management of its daily operations, an analytic tool developed by *Objective Logistics*. Basically, *Muse* tracks servers' performance over time based on strategic metrics such as sales per customer and customer satisfaction. Indeed, this enabled the chain to come up with a productivity-based ranking system whereby staff (waiters) have the possibility of choosing both their schedules and the tables they wish to serve. The rewards of making better-informed decisions are evident: empowering personnel, suppliers or customers, and amassing extensive information frequently involving outlays and complications. In this context, it worth noting the case of Amazon, which owes its early success to its drop-shipping model, which enabled it to offer over 1 million books while stocking an insignificant less-than-1% share of the bestselling titles. For the remainder, the company dispatched orders to book wholesalers and publishers, who subsequently shipped them straight to customers. In this state-of-the-art model, Amazon's network of partners was held responsible for the management of inventories. Adding that, they, the partners, not Amazon, had to assume the risk of carrying books without having any prior bits and pieces of the probable demand for them. Yet, because the risk was extensively allocated, all entities involved in the process succeeded to manage their own share of the risk with quite a good level of comfort.

– *Choose the decision-maker with the most to gain*

In several BMs, strategic decisions are often made by those with less to gain than others in the value chain. A company's customer, for instance, over and over again, feels they gain less when they purchase a company's product in comparison to when the company does. That was an issue fronting NETAFIM, a global leader in drip irrigation solutions for sustainable agriculture. By definition, smart drip and micro-irrigation is the watering technique of choice for individual farmers in tropical countries[22]. The company created a technology that tweaks water applications based on the soil's water content, salinity and fertilization, and according to climatological facts. NETAFIM exhibited to farmers that its technology could grow crop yields by 50%, and decrease water consumption by 40%, thus making it a hypothetically worthwhile investment[23].

The technology was unpopular at first; a tough sell indeed. Individual farmers were hesitant to accept the new, sophisticated technology and pay a price for it. They felt lots of risks would be eventually put on their shoulders if they decided to espouse it. NETAFIM overcame this problem by providing them with a free-of-charge, all-integrated package, including maintenance services. Reimbursement came from a share of each farmer's improved agricultural output. In other words, the company decided to take on all the risks of the decision, and farmers just had to respond to the opportunity with a *yes* or a *no*, with no possible downsides, to make extra money. The company succeeded in its mission because it understood that it had the most to gain from the acceptance of its technology. The risks bore by the company were a lot smaller than what they could have been if the small farmers were to assume them singlehanded. Also, even when the system botched their work with some farms, the company was able to make up for it elsewhere. And as individual farmers accomplished larger successes, through word-of-mouth, NETAFIM increased its sales figures and achieved economies of scale.

3.5.4.4. *What do key decision-makers do?*

Whenever decision-makers work together to create value, they should be able to fulfill their personal goals, while ensuring that the value chain won't be damaged in any way. There are three distinct ways to do this:

22 Netafim (2021). Drip Irrigation Changes the Face of Agriculture [Online]. Available at: https://www.netafim.com/en/drip-irrigation/ [Accessed June 15, 2021].

23 Permira (2021). NETAFIM ESG Case Study [Online]. Available at: https://www.permira.com/about/investing-responsibly/esg-case-studies/netafim/ [Accessed February 5, 2021].

– *Vary revenue streams*

Customarily, whenever the US DoD (Department of Defense) buys a jet, it would settle to a time-and-materials agreement, under which contractors charge a certain amount of money for labor and resources exhausted in due course of every maintenance event (similar to what mechanics do regarding car repairs!). Inopportunely, this BM did not endow contractors with client-friendly incentives. From their standpoint, the more the client faces difficulties, the better it is for them. And so, it has been projected that for every USD 1 the US DoD spends to buy a new jet, it spends an extra USD 7 on the aircraft's lifespan. Until, that is, the US DoD gave contractors a motive to care more about engine performance. In 2003, facing pressure to lessen costs and progress achievement, it embraced so-called performance-based contracting, which transformed the revenue model for contractors[24]. Under this BM, contractors are rewarded for the amount of time the jet is actually (or remains) in service. And so, the longer a jet performs without needing to be taken out of service for whatever reason, the more the contractors earn.

– *Orchestrate time horizons*

Usually, sourcing depended upon competitive-bidding schemes that guaranteed truncated prices and judicious, but satisfactory, quality services. The selected supplier won the business for a relatively short period of time, after which the bidding process was reinitiated. But as foreign sourcing amplified, the BM in use started to demonstrate some significant flaws. Distant suppliers began to cut corners on quality control and materials' trustworthiness. Even worse, disclosures of offensive labor practices, product diversions and the forging of goods materialized. And because most sourcing transactions consisted of on-off deals, careless suppliers remained unpunished – until, certainly, some multinationals started feeling the destructive impact of recurrent performance snags on the profitability of their brands.

Li & Fung, a Hong Kong-based logistics, commerce and distribution company, altered the world of outsourcing through the creation of a new BM, based on uniting the suppleness of sustainable sourcing with the sureness of durable rapports. It chooses, authenticates and accepts suppliers, and allots their businesses between its manufacturing clients. It also manages each client's relationship with each supplier; counting productivity, acquiescence and creating incentives for suppliers to invest in staff and resources. Given the prospective for a long-lasting rapport with Li & Fung, suppliers worldwide were stirred to create long-standing value for

---

24 Girotra, K. and Netessine, S. (2014). Four Paths to Business Model Innovation. *Harvard Business Review* [Online]. Available at: https://hbr.org/2014/07/four-paths-to-business-model-innovation [Accessed June 15, 2021].

manufacturing associates[25]. Yet, today, companies such as Li & Fung are rare, and to carry out sourcing transactions in sectors or industries where there is a significant lack of trusted intermediaries could be problematic for companies as they would be obliged to manage such rapports all by themselves, directly.

– Partner within an ecosystem of cross-industry actors

At present, prominent companies across industries have resolved that they could no more play it solo, contemplating ways for their businesses to magnify, transform – or just compete. Specifically, for companies to craft new opportunities or overthrow challenges, they need to build sustainable partnerships within an ecosystem of cross-industry players and work with partners to outline, shape and implement customer solutions.

In fact, executives across industries seem to be more and more interested in ecosystem-based BMs[26]. The ICT sector is among the earliest adopters of such systems. Indeed, tech giants were the first to acknowledge the opportunity that could ensue from bringing together application developers for the joint benefit of their platforms and customers. A brief overview of how this sector has leveraged ecosystems over time, we believe, would possibly help companies in other industries to efficaciously steer their own ecosystem strategies.

Customarily, technology players used to build ecosystems within their own industry galaxy by merging the competences and resources of other technology partners. Yet, they progressively stretched outside their core business by developing new businesses via the combination of skills and values conveyed by other industry expert partners. For instance, key ICT vendors, whose specialty was to develop products and sell them to infrastructure customers, ended up reaching out to services companies to cultivate and incorporate bespoke technology solutions for their new customers in the banking, health and public service sectors (among others). In this context, a recent study published by Accenture[27] showed that there is a substantial number of businesses around the globe (over 40%) that chose to construct their ecosystems outside of their own industry, wishing to explore new markets and serve new customers.

---

25 Donaldson, T. (2017). Here is How Li & Fung Plans to Create the Supply Chain of the Future [Online]. Available at: https://sourcingjournal.com/topics/business-news/heres-li-fung-plans-create-supply-chain-future-64351/ [Accessed June 15, 2021].

26 Artero, D. (2018). An ecosystem playbook for agility and growth [Online]. Available at: https://www.accenture.com/us-en/insights/strategy/ecosystem-playbook-agility-growth [Accessed June 15, 2021].

27 Lyman, M. (2018). Cornerstone of future growth: Ecosystems. Research report [Online]. Available at: https://www.accenture.com/us-en/insights/strategy/cornerstone-future-growth-ecosystems [Accessed June 15, 2021].

Put simply, joining an ecosystem is likely to provide businesses with an exclusive opportunity to drive innovation and lead disruption. Today, there is an enormous need to progress collective partnerships beyond customary associations and channel rapports. Thus, building an ecosystem, we proclaim, is a great move towards fostering disruption or, at least, responding to unwanted disruption. For example, in China, given that the app market for smart phones in the country is extremely fragmented, this opening has led to innovation intended to disrupt incumbent players. As a result, mini-programs were developed via a new ecosystem. Widely held by customers, they could be easily integrated into the instant messaging app and used effortlessly, without having to download them. And the pay-off? The disruptor gains give insight into customer behavior that could be used to create added value and to fortify the rapport with existing consumers. Besides, to preserve (or strengthen) their current positions in their core markets and sustain competitiveness, it is compulsory for businesses to stay attentive and keep on exploring (new) ways to work in collaboration with their partners. Actually, one of the prime threats dreaded by businesses today comes from new rivals outside of their industry[28]. We add that the latter would become lethal if businesses refrained from acting proactively and preventively.

On another note, Accenture's research indicated that about 50% of management teams, although in favor of ecosystem strategies, frequently see a risk in swapping information with fellow associates. To be fruitful, the rules of engagement among partners should be plainly voiced so as to curtail risk as much as possible, and to make sure each ecosystem partner contributes a unique role while upholding a basic level of control.

Ultimately, it is a trade-off, a give-and-take process between two companies, discerning the willingness of a company to share individual value versus the willingness of another company to gain joint value. For example, in the software application development market, tech giants swiftly figured out how to strike the right balance. They principally remunerated developers for their creations, their ability to build a critical store size, and also profited from any in-store purchasing transactions since they offered the general applications hosting the platform. And so, as revealed by technology tycoons, businesses should not, in any case, shy away from investing in, or adapting their business models, to reflect the sole needs of an ecosystem-based business strategy. This is because innovation, we affirm, could ensue in the creation of new markets, services and solutions – this could benefit quite a few companies, small and big, operating in various business sectors.

28 Ibid.

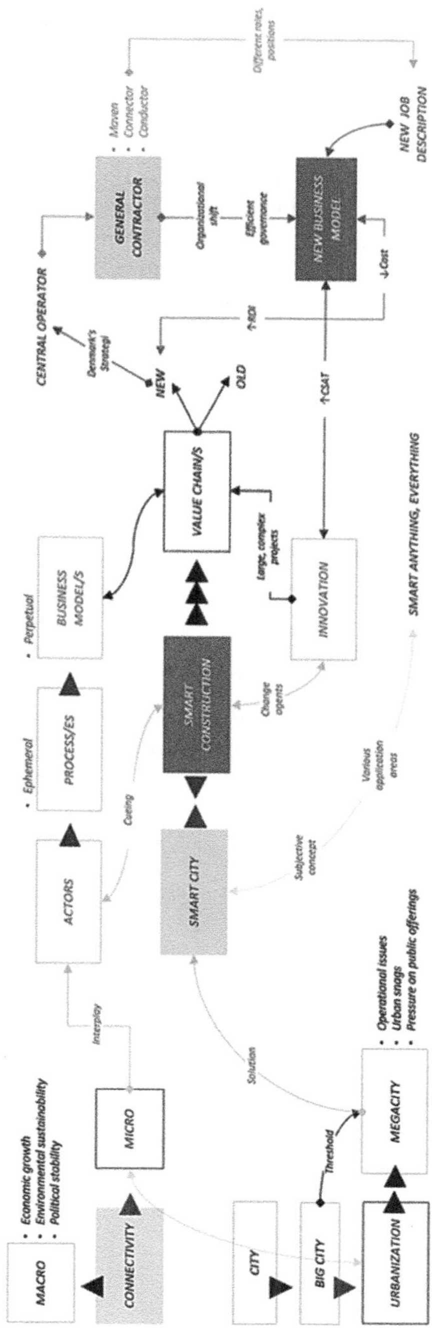

Figure 3.12. *Conceptual framework (source: created by the author). For a color version of this figure, see www.iste.co.uk/karam/general.zip*

3.6. From the managerial question to the research questions

Now that we have embarked upon the bulk of notions and concepts that are straightforwardly linked to our research topic in this book – going from *connectivity* and *urbanization*, to *smart cities* and *smart constructions*, to *innovation in construction* and *construction project management processes* and *value chains*, to finally reaching *BM theories* and *BMI* (Figure 3.12) – we proceeded in Chapter 3 to explain the various business modeling tools and research techniques exploited for the successful achievement of our empirical research. Concisely, we recap, our envisioned BM is solution-based and human-centric. As for the changes it is likely to bring to the table, those will be incremental rather than radical, of a managerial rather than technological nature. Moving forward, in Chapters 4 and 5, we suggest relevant answers to some interconnected managerial questions (and sub-questions) – some of which are as follows:

– How will the design of a new general contractor BM eventually help resolve the oh so many issues that construction companies are encountering in the management and overseeing of large, smart developments?

– How could such a BM help construction companies to better orchestrate the design, construction and delivery processes of smart city projects?

– How will it be designed? Based on which process? And what form will it eventually take? How many service layers will it comprise?

– What will be the role of the *general contractor* in particular, in comparison to other key construction players? What about end-users and ICT companies?

4

The Design Process of the General Contractor Business Model

Nowadays, cities around the globe are being put under increased pressure to host more and more inhabitants. The influx of people towards big cities has indeed become a trend as people are always looking for better opportunities and enhanced connectivity. It is the attractiveness of cities, their economic power and influence that would depict whether they would eventually be regarded as hotspots or not. Urbanization has actually set the new rules of the game, and cities were put to the test as a result. Essentially, there are two plausible paths that cities could undertake: adapt to the new realities and flourish, or fail to do so and lag behind. Apart from urbanization, technological developments have also led to the world, including the ways we live, communicate with each other and do business, evolving constantly – at an accelerated pace. Since the Internet bubble in 2001, the world seems to be in a race against itself. Companies are persistently searching to exploit the best technologies available in order to improve competitiveness and productivity, sustain their businesses and secure a competitive edge over their rivals.

Alike companies, cities too, are currently seeking to strengthen their respective market standings. Their overall appeal is at stake. And so, they are presently in a race to attract more resources, talents and investments. The only means for them to win the race is by becoming smart. Hence, we proclaim, cities, mostly big ones, are destined to become smart in order to survive the many urban challenges they are facing today. Thus far, it is apparent that some cities have done a better job in this respect than others. While Barcelona, London and Copenhagen are now allegedly smart, France, for instance, seems to still be figuring out how to catch up with its neighboring metropolises. On the other side of the world, in Southeast Asia specifically, Singapore constitutes a great example of a city that failed to keep pace with development at first, but eventually succeeded and took the lead. We hence look back at a question that has been raised in the earlier chapters of the study, namely: *have all cities vowed to be smart?* As put by Mr. C.R. (professor at a well-known University in Paris),

> Lee Kuan Yew, the founder of Singapore, succeeded in building the city of tomorrow by simply copying other successful smart city models. While imitation was the primary component of his strategy, he also made sure to integrate the state's own ideologies and values into it. The whole system he created was founded on Singapore's own cultural tradition.

This puts emphasis on the fact that a smart city should not be disjointedly conceived from the values and cultures of the territory on which it is sited, but on the contrary, it should be a direct reflection of them. According to Mr. C.R.,

> The French model is fairly comparable to the one in Russia. In Russia the problem lies in the cultural framework of the country, that is, the prevalence of monocities, a legacy of a top-down (Soviet) conception of power and development.

Under the Soviet governance model, we proclaim that it is the government's role to bring happiness to the people, with the people having no clear role in this respect. Indeed, the social structure of the country is problematic, being a source of both corruption and bureaucracy, especially as the government often seeks its own interests and tends to stick its nose into every aspect of people's lives, which means that people have no other choice but to bargain with it in order to get what they need.

Following the same reasoning, Mr. V.B. (a program director at a well-known university in Paris) affirmed that "people are often inclined to crowd into cities". To further elucidate his viewpoint on the matter, he advanced the case of Chinese cities being centralized administrative economies with hundreds of millions of residents. Consequently, Mr. V.B. asserted that:

> Smart city initiatives are more likely to succeed in China than anywhere else in the world because the recipe of success is time and time again set by a single central entity. In Anglo-Saxon democracies decentralized decision-making schemes are privileged and the masses are often consigned to choose, which is not a bad thing *per se* – but

how could we eject all of the paradigms that may arise from the fact that we do not have centrally planned economies, but rather plentiful agents who operate based on short-term-yield business logics?

It is within this rationality that the idea of a GC BM for the building of smart cities sprang to mind. We herein explicate how the latter materialized and was nurtured over time. And also how it gradually, little by little, transformed into a tangible artifact that, we hope, could serve construction companies in the execution and management of their future smart city projects. The four-phase process on which this new BM idea is based is illustrated in Figure 4.1. As for the main data sources used per phase, they are listed in Table 4.1.

PHASE 1. RISE	*PHASE 2. SHOWCASE*	*PHASE 3. MATURITY*	*PHASE 4. EVALUATION*
– Own professional experience	*– Own professional experience*	*– Meeting with prominent industry players*	*– Meeting with prominent industry players*
– Brushing upon academic literature	*– Revising and examining construction industry publications*	*– Partaking in collaborative research efforts on the smart city topic*	*– Meeting with scholars and academics*
– Meeting with prominent industry players	*– Meeting with scholars and academics*		
– Attending smart city conferences and exhibitions	*– Meeting with prominent industry players*		

Figure 4.1. *The development process of a GC BM for the building of smart cities (source: created by the author). For a color version of this figure, see www.iste.co.uk/karam/general.zip*

Within each phase of the process, problems were identified and solutions were suggested for their resolution. A thorough explanation of the problems and solutions is available in Chapter 5.

PHASE 1. RISE	*PHASE 2. SHOWCASE*	*PHASE 3. MATURITY*	*PHASE 4. EVALUATION*
Main Readings: *– Flyvbjerg (2014)* *– Virtanen et al. (2014)* ***Conferences & Exhibitions:*** *– The smart city and smart grids exhibition, October 4-5, 2017 (Paris porte de Versailles)* *– Forum Smart City Du Grand Paris 2017 – La Tribune (Vivre Ensemble) - November, 2017 - Hôtel de ville de Paris*	***Main Readings:*** *– Flyvbjerg and Holm (2002)* *– Flyvbjerg et al. (2003)* *– Le Moniteur (Batiactu)* *– Industry Consultancy Reports (Bouygues, Vinci and others)* ***Conferences & Exhibitions:*** *– The smart city and smart grids exhibition, October, 2018 (Paris porte de Versailles)* *– Forum Smart City Du Grand Paris 2018 – La Tribune (Vivre Ensemble) - November, 2018 - Hôtel de ville de Paris*	***Main Readings:*** *– Industry reports (Eiffage, Bouygues, Vinci and others)* ***Conferences & Exhibitions:*** *– Conferences given by Architects at École Nationale d'Architecture Paris-Val de Seine (ENSAPVS)* *– The smart city and smart grids exhibition, October, 2019 (Paris porte de Versailles)*	*Not Applicable. The validity of our business model was conversed and evaluated by prominent industry players and academics.*

Table 4.1. *Data sources used in the business model design process (source: created by the author). For a color version of this table, see www.iste.co.uk/karam/general.zip*

4.1. Phase 1: rise

Over the course of my professional career as a construction expert, working in the French construction industry for over 10 years, I have had the chance to accumulate sufficient knowledge about construction practices and procedures operated in France. Everything held constant – or in other words, for an external observer – the construction industry in France seems to be functioning just fine (which is true in general!), yet when it comes to smart constructions specificly, it certainly is not.

Just a few years back, the smart city concept started to grow in popularity and advance. It eventually turned into a global trend. Consequently, I, myself, being passionate about the topic, thought of creating a research-and-development department within the general contracting company I own to better understand the question of smart cities. Truthfully, it is not just the smart city concept that drew my attention, but also the current structure of the construction industry's value chain. Indeed, my aim was to figure out whether the latter is resourcefully designed to allow for companies to undertake smart construction projects and ensure their success in the long run. While being acquainted with the construction industry in Anglo-Saxon economies, I even questioned whether a construction company like the one I am heading could be integrated into the chain and be part of the solution to the various problems France-based construction companies are currently facing.

In France, the construction industry is highly fragmented and state-controlled, so suggesting something new and inciting construction actors to espouse it would not be an easy task to fulfill. I am fully aware of that. Nonetheless, the idea that I had in mind was vital and so, I assumed that it was worth the effort.

At the beginning, I believed that the main problem that was preventing the construction industry from flourishing was the existing mismatch between supply and demand. And the solution, I alleged, may simply take the form of a more adequate offering that could better fit the real needs and aspirations of end-users. Although my solution was to some extent legitimate, I suspected that a more detailed investigation of the topic would be necessary so that I could possibly come up with a global solution that could benefit not just my own company, but the entire construction industry. For that reason, I made the decision to investigate the smart city concept further and endow my research with some scientific credibility. To say the least, the decision I made enabled me to focus my efforts and work on the topic in question, and come up with a proposition that holds value in its layers.

Indeed, my research aimed to identify problems hindering the functioning of the French construction industry and suggest plausible solutions that could set it back on the right path for growth. As previously noted, urbanization is putting intense pressure on cities to innovate and magnify their appeal. The same goes for construction companies, who are obliged to rethink their offerings, so they better match with current market requirements. In France, at least to date, things do not appear to be rolling in the right direction. The inadequate BM on which the French construction industry is now operating has been significantly affecting the latter's performance and productivity. In other words, this implies that a change must be implemented in the industry's current value chain, so that it can better cope with shifting market realities and redress its economic figures. The suggested BMI revolves around a GC to be integrated into the industry's value chain, in an endeavor to bridge the various gaps found along it.

I interviewed various individuals (e.g. architecture firms, multidisciplinary design offices, smart city experts, public developers and others) to gather practical insights into topics discussed openly regarding the innovation I am proposing. Nevertheless, before I reached the interviews, I first did some exploratory research on the topics of smart cities, BMs, and GCs, in order to draft a discussion guide on which my discourse with interviewees would be structured.

The questions (Q1, Q2, etc.) listed below are only *headlines* that were used to direct the *interviewer-interviewee* discourse. The exchanges we had with interviewees were very broad, covering copious matters that are in direct or indirect association with our managerial questions and research topic.
All or most IDIs were conducted in French.

On the need for and design of a new BM:

- **Q1:** With regards to the management process of construction projects in general, what are the principal, most frequent flaws that construction companies should pay the most attention to in their day-to-day activities?
- **Q2:** What are the recurrent issues you used to face on a daily basis (whether it is your company or other companies you used to work with)? What experience could you draw from?
- **Q3:** What kind of improvements would you suggest, technological or managerial? What about the tools you are using? What are your suggestions for companies to be able to overcome these flaws and bridge the prevailing resources gap?
- **Q4:** Would the design of a new BM (reshaped, reinvented, or simply adjusted) help companies to solve these problems? If not, what other options do we have?
- How do the old and future value chains in construction compare? Where do the deviations (improvements) lie?
- **Q5:** Construction projects are intricate, with lots of information flowing and being exchanged among players during conception and execution: what do you think would be the best means to manage such tremendous data flows? Please explain.
- **Q6:** What do you think of smart cities? How would you define a smart city? Is it a real opportunity that construction companies should seize (in partnership with other entities)?
- **Q7:** What is the best way (blend of strategies and tactics) for companies to step into the smart city market? What does it entail? What are the best practices in this domain?
- **Q8:** With their current operating models (or business models), would French construction companies be able to embed and compete in the emerging smart city market? If not, what changes are required? Please elaborate.

- **Q9:** Do you believe operational and other benefits could be drawn from GCs handling (or leading) smart construction projects, from start to finish? Please specify. What would be the roles of other stakeholders? Under similar settings, how would the construction process run?

On the legitimacy, validity and sustainability of our envisioned BM:

- **Q1:** What do you think of our BM (design- and content-wise)? How does it compare to other original BMs that you are familiar with?
- **Q2:** What perfections or improvements do you propose (if any)?
- **Q3:** Are there any nuances that you would like to shed light on and that we might have neglected during the BM process design?

Box 4.1

The several readings I did, referring to both academic and nonacademic publications, somewhat confirmed my suspicions that the industry's BM is dysfunctional and that, consequently, some incremental radical changes should be brought to it. Moreover, they actually helped me to draft a review of the literature on the topics in question. Whether based on personal intuitions or insights gathered through interviews, all data pointed towards one thing: the need for a new entity to be consigned for the running and management of large construction projects, from initiation to closure.

My idea of a new GC BM was originally inspired by Flyvbjerg (2014), who put emphasis on the need for a new entrepreneurial entity to conduct lengthy megaprojects. Instead of thin slicing projects into numerous subprojects and executing them as part of disconnected arrangements, the authors proposed that the projects should be bundled into one contract, where a private sector entity would be commissioned to oversee and ensure the management of all works required. Furthermore, as mentioned by Virtanen *et al.* (2014), introducing a central operator into the old construction value chain would result in improved control over the entire construction process in terms of both scheduling and price determination. In addition, the authors avowed that the amount of logistics and pieces held in storage would be condensed and the subordinate quantities of resources that are tied up in the process would be freed. For end-users, they said that the amount of customizability would be significantly amplified.

In the interviews, an architect (whose identity remains concealed) who works in a well renowned, award winning, Île-de-France-based architecture cabinet affirmed that the French construction sector, apart from being fragmented, has now reached a certain level of maturity, where most CVP made are monotonous and outdated. "Innovation is required", he said, "not essentially technical or technological, but organizational".

In the context of smart city development, France is still trailing behind and something new is needed for it to catch up. The smart city market's potential, we proclaim, is huge and France should adjust its ways of doing, so that it can successfully secure its share of it. Largely, the interviewee showed lots of interest in the idea of developing a GC BM for the creation of smart cities. This has comforted me in the sense that the solution that I am trying to craft would crack an issue that several construction actors believe exists – not only myself. Another interview I conducted with the COO of a Multidisciplinary Design Office based in Île-de-France revealed fairly similar conclusions. Precisely, it has been agreed that although the smart city concept is vague and not very well defined in France, such an innovation, at least for the time being, is interesting, as it constitutes an actionable solution to the daily issues faced by the industry's stakeholders. For the COO:

> Though interesting and worth testing, it is hard to believe that such a role would be able to solve all of the problems encountered in the construction industry today. Such an innovation could, however, be part of a bigger, all-inclusive solution.

Another industry expert who was approached and shared their viewpoints vis-à-vis my BMI proposition was a large public developer who has been in charge of implementing several large construction projects in the Île-de-France area. He specialized in the development of eco-quartiers, eco-friendly constructions that are not necessarily smart. After discussing the foundations of his company and the mix of construction processes he usually adopts for the implementation of large projects, he openly asserted that there is a tendency today among public developers to delegate the ownership of their projects to private entities who have the required knowledge and expertise in handling and managing smart developments. This is, we proclaim, a change that pubic developers are in due course of bringing to their BMs.

As reported, although large projects are being successfully built and delivered, they often fall short in terms of meeting end-users' expectations. Indeed, there is a huge gap between what is being offered today and what end-users are actually expecting. This problem, we proclaim, has nothing to do with companies' construction processes, but rather their BMs. Time and again, end-users are dissatisfied with the quality of construction works and complain about the state of the final product received. This situation is further aggravated by responsibility diffusion, where, post-delivery of final product, stakeholders claim to not be liable for final-product state, and end-users see themselves struggling to find the entity who is (should be!) in charge of maintenance or after-sale works (e.g. insurance companies and property managers). For the interviewee:

> these problems could not be solved unless potential end-users and local communities become partakers in large construction projects. We are now trying to change our BM so that we do not operate in a pure economic-logic anymore.

Indeed, we say, the spectrum of construction projects has been changing recently, widening to cover the social and the environmental, as much as the economic. The interviewee said:

> Construction companies today should worry less about the economics of their projects and more about the well-being of citizens. Nowadays, we are investing both time and effort to revise and widen the scope of our BMs so that we can adjust our value propositions to the satisfaction of end-users. If we cannot keep the promise we make to end-users at the start of the project, then there is definitely something wrong in the BM that we are operating on.

Often, value created in the very early stages of the production chain is lost in the later – more advanced – stages. This difficulty, we aver, could be resolved once a specialized entity has been consigned to oversee and keep track of all works performed at each construction stage. In conclusion, the smart city conferences and exhibitions I attended – mainly the Smart City & Smart Grids Exhibition and the Forum Smart City du Grand Paris – were useful as they have given further meaning to my BM proposition. Indeed, they clarified the nuance that smart city projects should not in any way solely revolve around the technical and technological. As communicated, gadgetizing our cities would not inevitably make them more attractive to people, but on the contrary, we risk making them less humane. We intend to make our cities smarter by reinventing them, and reinventing them is not a simple question of installing smart grids (among others) at city level, but allowing all builds in the city to create a coherent and resilient system where data freely flows from one build to another. The 21st-century challenges (e.g. climate change, energy efficiency and reduced carbon footprint) are numerous, and our cities need to constantly evolve to be able to cope with them. One moderator said:

> Reinvesting Paris means summoning all of the creative forces to continue to imagine a city moving ahead in its metropolitan constructions; it is to give our city an original urban radiance so it remains an international attraction; also, it is about imagining new ways for the city to create more wealth and jobs, and encourage data sharing and knowledge exchange among its constituents.

Table 4.2 exhibits the main problems identified in phase one, as well as the solutions suggested for their resolution. Accordingly, a first draft version of our new BM has been envisioned (see Figure 4.2). It accounts for four variables, explicitly: smart city components, construction stages, construction actors and dimensions of a construction project. The color codes were carefully chosen to show inherent similarities between the dimensions of a construction project, on the one hand, and the components of a smart city, on the other hand. Moreover, our choice of variables ensued from our understanding that construction projects are stage-based and that construction works evolve every time we move from one stage to the other. In parallel, we presume that construction actors intervening along the process would change, with one stepping in and another stepping out at each construction stage. Therefore, we say, the execution of construction projects follows a systematic course where the relay, stage after stage, is passed on from one stakeholder to another, and it is their collective effort that would finally result in the building of smart constructions. The dynamics behind our suggested BM are shown in Chapter 5, using a specialized software that was developed to this particular end.

<table>
<tr><th>Identified problems</th><th>Suggested solutions</th></tr>
<tr><td>– Urbanization</td><td>– (Convert big city into smart city)
– Improve attractivity and absorption capacity of city</td></tr>
<tr><td>– Failing BM leading to abridged growth prospects at industry level</td><td rowspan="2">– Rely on organizational change to drive industry growth</td></tr>
<tr><td>– Project process confused for business model</td></tr>
<tr><td>– Outdated CVP</td><td>– Adjust offerings to the exact needs and requirements of end-users
– Allow for customizability (portfolio diversification)</td></tr>
<tr><td>– Constricted approach to value creation</td><td>– Ensure final product has a use-value and that utility may be extracted from it</td></tr>
<tr><td>– Misconception of smart city notion</td><td>– Partner with local communities for the conception and building of smart developments</td></tr>
<tr><td>– Smart city projects as disconnected micro projects</td><td rowspan="5">– Rely on organizational change to drive industry growth</td></tr>
<tr><td>– Fragmented projects</td></tr>
<tr><td>– Too much weight put on technology in the building of smart cities</td></tr>
<tr><td>– Smart constructions as pure technology demonstrators</td></tr>
<tr><td>– Overlooked organizational innovation</td></tr>
<tr><td>– Large amount of logistics, pieces and other resources tied in the process</td><td>– Abridge logistics and procurement costs
– Assign central operator to oversee execution works</td></tr>
</table>

<table>
<tr><td>– Loss of control over quality of works</td><td>– Assign central operator to oversee execution
– Set cohesive plan for the execution and tracking of construction works</td></tr>
<tr><td>– Mismatch between promised and delivered final product</td><td rowspan="2">– Partner with local communities for the conception and building of smart developments
– Organize hearing sessions for all people affected by the project
– Assign central operator to oversee execution works
– Assign specialized entity to handle after-sales services and maintenance works</td></tr>
<tr><td>– Criticism of final product by end-users</td></tr>
</table>

Table 4.2. *Problems versus solutions (phase 1) (source: created by the author)*

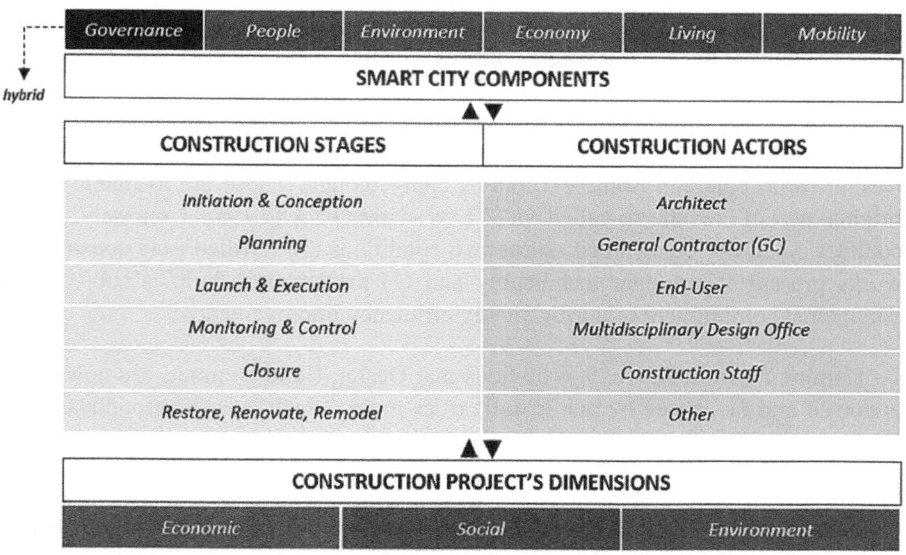

Figure 4.2. *Version 1 of GC BM (source: created by the author). For a color version of this figure, see www.iste.co.uk/karam/general.zip*

4.2. Phase 2: showcase

In phase 2, after capitalizing on the findings of phase 1, we made a step forward in the direction of building a more comprehensible GC BM for the creation of smart cities. Affirmatively, the exploratory research we did and the findings we gathered gradually nurtured – phase after phase. Further problems were identified and plausible solutions were suggested during this phase.

After brushing up on the industry's most relevant publications, we noticed that among the various challenges that construction companies face, lay state-imposed construction standards and codes. They are being repeatedly revised at an accelerated pace[1]. For construction companies, such tweaks in regulations often ensue in *delays in project execution*, as on-going projects must be stopped and necessary changes made before construction could resume. Apparently, today, neither construction companies nor the people are able to keep track of those supervisory innovations.

Within this setting, we add some recent texts in construction law (e.g. Élan Law, Article 18, relating to Design-Build contracts)[2] that seem to favor some specific stakeholders and disfavor others. Largely, some construction actors, mainly architects, are now feeling threatened by such regulatory innovations as the odds of their roles being marginalized are high[3]. Following the Élan Law, the project owner now has the freedom to deviate from the principle of state law stating that the tasks of the PM and the GC are necessarily distinct. Thus, both tasks could be (alternatively) assigned to a single entity who could concomitantly play both roles. In other words, the project owner may override the process of awarding two separate contracts and entrust the design and the execution of a project either to a group of operators (the so-called in French "ensemblier") – or to a single private entity via a Design-Build contract. Some exceptions apply though. Those are available in the original text of law, the so-called MOP Law, dating back to 1985. Concisely, such a contract could not be awarded unless two conditions are fulfilled[4]: (1) construction works present a particular technicality and (2) the structure built is subject to a contractual commitment requiring energy efficiency improvement[5].

Continuing from above, it is obvious that Design-Build contracts are now being endorsed and facilitated by public authorities more, which lays the groundwork for an innovation like the one we are suggesting in this study. Moreover, it is practically

1 Juridique Les contrats globaux de construction en Europe. Available at: https://www.lemoniteur.fr/article/juridique-les-contrats-globaux-de-construction-en-europe.45309 [Accessed June 15, 2021].

2 CONSTRUIRE EN CONTRAT GLOBAL. Available at: http://boutique.lemoniteur.fr/construire-en-contrat-global.html#caracteristiques_produit [Accessed June 15, 2021].

3 Les contrats globaux, seul moyen de construire les bâtiments de demain. Available at: https://www.batiactu.com/edito/ontrats-globaux-55693.php?MD5email=3bda47a71aa107686b2e568cde6824a1&utm_source=news_actu&utm_medium=edito&utm_content=article [Accessed June 15, 2021].

4 Juridique Les contrats globaux de construction en Europe. Available at: https://www.lemoniteur.fr/article/juridique-les-contrats-globaux-de-construction-en-europe.45309 [Accessed June 15, 2021].

5 Associated with the Grenelle II law of 2010.

conceivable to expect the management and execution of large construction projects to change (for the better) in the near future. Within the realm of contracts, we carry on by stating that big operators, such as Bouygues and Eiffage, have recently used global contracts for their own benefit. Actually, whenever a large construction project is commissioned, big operators tend to subcontract, through passive bidding processes[6], some or all of the work to third parties. Subcontractors selected for the job are often vulnerable and set to work at near-zero margins. In other words, big operators frequently outsource construction works to the lowest bidders, so that they can keep high margins for themselves. This constricted approach to value creation is one of the main problems that the construction industry is currently facing. At times, big operators may possibly stop the project without even paying the subcontractors. They tend to do so whenever the project is thought to be economically unfeasible, which supervenes in operational issues that accumulate throughout the process to finally reach the end-user. This could partly explain why end-users are often dissatisfied with the final products they get – often being of mediocre quality, unlike the products promised at the start. And so, big operators tend to neglect the well-being of end-users over and over again in a short-sighted attempt to maximize their own profits.

Following a meeting I had with the Chief of Innovation Services at a Multidisciplinary Design Office based in Paris, we agreed that there is an actual need for an innovation at industry level, and that the institution of a new organizational structure for the management of large complex projects is highly recommended. As put by the claimant, as customers' needs are constantly changing, so should our offerings, otherwise our company will lag behind and see its customer base shrink.

Although most companies would find it difficult to rethink their BMs, they must come up with an innovative strategy that would enable them to redress their market standing and remain competitive in the marketplace: our construction processes are impeccable, yet what is outdated is our customer value propositions.

By focusing on the economics of construction projects, we often neglect the social and environmental. The economics of a project are only worthwhile for construction companies, but are worthless for customers. What customers really care about is the social and environmental aspects of projects, which are so often ignored. According to a Business Development Manager who works at a Multidisciplinary Design Office in Paris:

---

6 This sheds light on the fact that operational problems encountered during the conception and execution of smart developments originate before the kickoff of construction projects – in the course of bidding processes.

> A smart city does not exist in France – not yet at least. It is the prevalence of short-term financial goals among construction companies that renders our constructions rigid and inefficient. Big operators only care about themselves, their own profits, and do not care about the impact of their projects on people and the environment.

Another architect – specializing in the construction of smart residential structures for the elderly – with whom I had the chance to meet and speak, shared the same viewpoint in this respect, stating that such an innovation (i.e. a GC BM for the building of smart cities) is definitely required: first to palliate the different problems construction companies are currently facing, second to allow France to step into the international (huge) smart city market, and third to ensure that what we are delivering to end-users goes beyond simple walls and roads. The architect said:

> Apparently, today, construction companies, big ones in particular, are working on hedging their portfolios by rethinking their own BMs – or say, by extending their scope of work so they can better satisfy all market needs and requirements.

Indeed, some France-based construction companies are currently seeking to espouse the role of an integrator by trying to extend their business activities, in an endeavor to cover the entire value chain, down to upstream[7]. For these companies, any attempt to conceive the city of tomorrow without the help of local communities would be destined to fail. Thus, we affirm that a smart city cannot be built without a successful collaboration between all stakeholders immersed in the project, including end-users.

Hence, under this vision, end-users should become active construction actors who co-conceive and co-design smart projects along with construction companies. To this end, organizing public hearing sessions (concessions) with local communities and potential end-users is a stepping-stone to allow end-users to start playing a more active role in the conception and building of smart cities, and boost social acceptability for the smart structures built. Shockingly, it is only recently that big operators realized that conceiving smart cities following a top-down approach would fail. They saw their financial earns drop and their stocks of unsold units grow. And so, they realized that smart cities cannot be built without cooperation with end-users, following a bottom-up approach.

7 To my surprise, during my investigations, I found out that insurance companies in France, some of them at least (AXA), have a general contractor agreement that allows construction companies to expand their business activities downstream the value chain. However, under such a contract, a general contracting company, although overseeing and controlling the entire construction process, is bound to outsource all trade to third parties – they are not permitted to carry out any construction by themselves.

Within this framework – and according to the General Manager of a financial company whose activities are focused on social housing – the GC role may only be consigned by the project owner (or big private operators). A maven in construction, they would play the role of a delegated project owner, replacing the often-deemed-incompetent original project owner throughout the entire project lifecycle, and be held responsible for the success or failure of the project. They would oversee all construction works carried out and synchronize the tasks of different actors involved in the process.

The diversity of stakeholders (e.g. architects, engineers, urban planners, contractors and subcontractors) involved in construction projects is not a problem *per se* – the interviewee signposted – but on the contrary, it is a healthy thing, yet what is needed today is someone, an entity that can orchestrate and direct their works, to ensure that they are being achieved harmoniously rather than separately, and ensure that the project is delivered on time and within budget. This is important as faulty estimates are often seen as a key characteristic of large construction projects – a problem that often leads to budget overruns, thus pushing construction companies to adopt cost-cutting measures in an endeavor to meet their project's budget. Such measures, we proclaim, habitually influence the social and environmental rather than economic aspects of the project. This is true as construction companies are less inclined to sacrifice part of their economic returns to uphold the agreed upon social-and-environmental specificities of the project.

Moving forward, during my conversation with Mr. V.B., a program director at a well-known university in Paris, the topic of social housing was raised. In social housing (relative to private housing), citizens are at the core of the BM. In Mr. V.B.'s words:

> The viability of the social housing model is greater than that of private housing because the former is maintained and preserved via public governance schemes, by which the life span of properties could be distended. The social housing model, reinforced by a smart governance structure within which citizens are key actors, may be adopted for the conception and building of smart cities. Such a model would ensure co-conceived constructions are accessible and affordable for all.

For construction companies – mainly private ones – social housing remains an underserved market, an opportunity they should seize in their furious attempt to build smart cities. In conclusion, a university professor specialized in issues of socioeconomic and territorial changes, whom I met during the Smart City's Exhibition held in Paris, insisted on the fact that the successful building and running of smart cities is primarily a question of governance and financing, hence the need for both central operators who could bring all the constituents of a smart city together, and investors who could procure the monetary funds required for the building of the city:

> Unless there is a certain form of interoperability between the different components of a city, the latter would not in any way be deemed smart.

<table>
<tr><th>Identified problems</th><th>Suggested solutions</th></tr>
<tr><td>– Passive bidding processes</td><td rowspan="4">External (uncontrollable) factors</td></tr>
<tr><td>– Legal and institutional barriers</td></tr>
<tr><td>– Constantly changing construction standards</td></tr>
<tr><td>– Financing</td></tr>
<tr><td>– Failing BM leading to abridged growth prospects at industry level</td><td>– Identify and target underserved market segments
– Allow for customizability (portfolio diversification)
– Adjust offerings to match the exact needs and requirements of end-users
– Ensure that the final product has a use-value and that utility may be extracted from it</td></tr>
<tr><td>– Big companies' resistance to change</td><td>– Rely on organizational change to drive industry growth</td></tr>
<tr><td>– Top-down managerial approach</td><td>– Administer smart developments following a bottom-up managerial approach</td></tr>
<tr><td>– Banishment of any collaborative urban governance scheme</td><td>– Rely on collective thinking and intelligence to conceive and execute smart developments
– Promote collaborative urban governance schemes (social constructivism)
– Partner with local communities for the conception and building of smart developments
– Create value in networks – profit sharing logic
– Rely on services of external partners to achieve specialized tasks (outsourcing)
– Build mutually beneficial rapports with stakeholders</td></tr>
<tr><td>– End-users not being co-creators of value</td><td>– Promote social engagement of end-users in construction projects
– Co-create value by co-conceiving construction projects
– Ensure the involvement of end-users in all stages of the construction process</td></tr>
<tr><td rowspan="2">– End-users not involved in the early stages of construction</td><td>– Organize hearing sessions for all people affected by the project</td></tr>
<tr><td>– Recognize end-users as both data sources and innovation creators</td></tr>
<tr><td>– Faulty estimates leading to recurrent budget overruns</td><td>– Set cohesive plan for the execution and tracking of construction works</td></tr>
<tr><td>– Scheduled timelines left in tatters</td><td>– Set efficient budgeting systems and flexible timelines</td></tr>
</table>

<table>
<tr><th>Identified problems</th><th>Suggested solutions</th></tr>
<tr><td>– High number of stakeholders with overlapping roles and responsibilities</td><td>– Espouse a centralized decision making process
– Manage stakeholders involved in the process and clear up roles and responsibilities</td></tr>
<tr><td>– Discoordination between construction actors</td><td>– Assign a central operator to oversee execution works
– Set up effective stakeholder management strategies</td></tr>
<tr><td>– Decentralized decision-making</td><td>– Ensure apposite cueing and synchronization among construction actors
– Set up effective stakeholder management strategies
– Build a lengthy network of industry partners</td></tr>
<tr><td>– Absence of an entity that could orchestrate time horizons and handle maintenance works</td><td>– Assign a specialized entity to handle maintenance works
– Responsabilize stakeholders
– Assign a central operator to oversee execution works</td></tr>
<tr><td>– Lack of understanding of smart construction process</td><td>– Promote social engagement of end-users in construction projects
– Administer smart developments following a bottom-up managerial approach</td></tr>
<tr><td>– Risk of technological obsolescence upon delivery of project</td><td rowspan="3">– Set cohesive plan for the execution and tracking of construction works
– Better innovation management practices
– Lower risk of premature technology obsolescence
– Improve work quality and curb maintenance costs</td></tr>
<tr><td>– Additional costs incurred to integrate new technology and upgrade asset</td></tr>
<tr><td>– Innovation mismanagement</td></tr>
</table>

Table 4.3. *Problems versus solutions (phase 2) (source: created by the author)*

Per se, a smart city is a platform or an ecosystem where all of its builds communicate with one another. Communication, he added, is often facilitated through technology. In addition, by allowing the free flow of data within the city, people could use it to come up with innovative ideas that could be exploited to further enhance the city's performance and appeal. Therefore, we affirm, building a smart city is an ongoing process that is subject to continuous improvements. Thus, a BM that is specifically catered for smart cities should be flexible and adaptable in the sense that it should fit in to any improvements made.

Table 4.3 illustrates the main problems identified in phase 2 and the solutions suggested for their resolve. Here, we wish to reiterate that the problems and solutions listed are only headlines and will be explored in further detail in Chapter 5. Accordingly, a second draft of the BM has been created (see Figure 4.3).

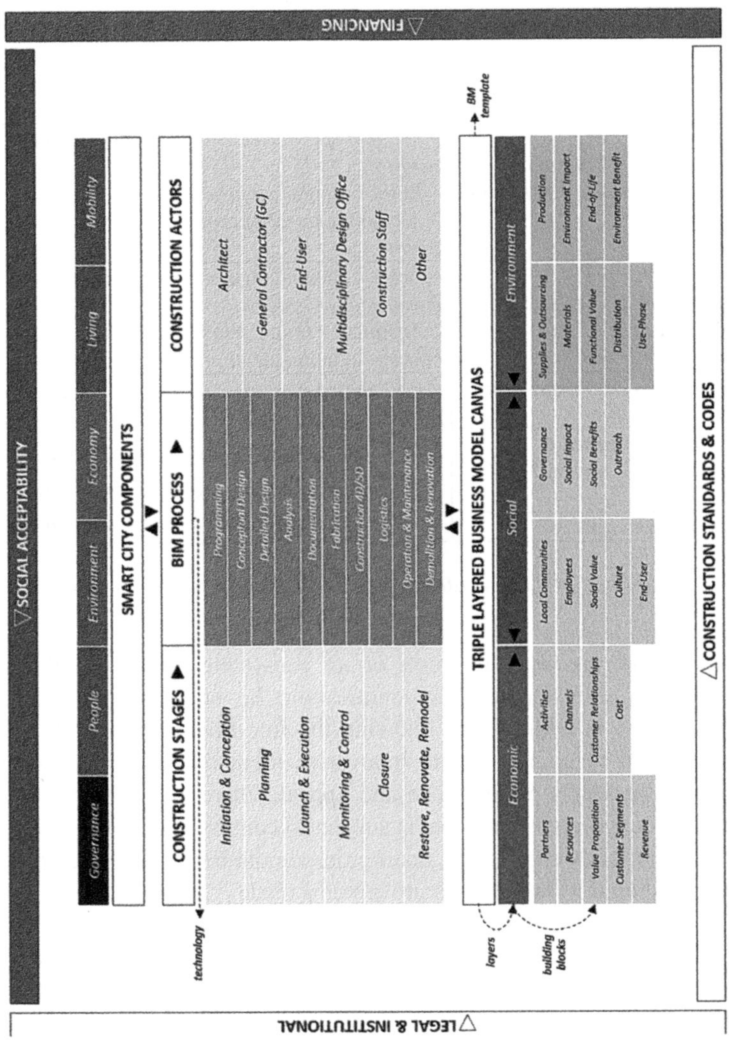

Figure 4.3. *Version 2 of GC BM (source: created by the author). For a color version of this figure, see www.iste.co.uk/karam/general.zip*

As is evident, some significant improvements have been made to the first version of our proposed BM – with some variables added and others replaced in version two:

– The TLBMC, being a strategic management tool, has been opted for so we could better reflect the various dimensions of a construction project, namely: economic, social and environmental. In the TLBMC, dimensions (version 1 of BM) are rather connoted layers (version 2 of BM). And within each layer, corresponding building blocks were added. The TLBMC allowed us to narrow down our problem-solution analysis, and served as a scheme for the portrayal of our GC BM. In other words, the various problems identified and solutions suggested at each phase of the BM design process would ultimately be allotted across the numerous builds of the TLBMC.

– Given that innovation is often enabled by technology, we chose to add the BIM (Building Information Modelling) Process into our BM, a tool that GCs, in partnership with other key stakeholders, could use to better plan, design, build and manage large construction projects.

– Finally, we drew the boundaries of our GC BM for the creation of smart cities. Indeed, four variables were identified, explicitly: (1) legal and institutional aspects; (2) construction standards and codes; (3) financing; and (4) social acceptability. Indeed, we say, any construction project, smart or not, cannot be executed unless it complies with state-imposed rules and regulations – including construction codes and standards; it cannot be built unless the required funds are secured first, and, finally, cannot be sold unless requested and accepted by end-users.

4.3. Phase 3: maturity

In phase 3, we affirm that our envisioned GC BM has reached a certain level of maturity and is somehow complete. We are not proclaiming that our GC BM is perfect, not at all. It is simply a serious attempt in the direction of creating an artifact that construction companies in France (or anywhere in the world) could use to better tackle the question of smart cities.

Ensuing from our findings in phase 2, I carried on my investigation by fine-tuning the second version of the GC BM. Based on a meeting I had with the Business Development Manager at a large construction corporation, I was able to gather some invaluable insights about customers' real needs, mainly what they want and what they aspire to get. To quote the interviewee: "a sociologist once said, what people actually want is to feel like they are permanently on vacation". Although they live in the city, the village must be brought to them – hence, the importance of encouraging smart initiatives like the Chinese made urban-village. Besides walls and roads, construction companies must look at construction from a different perspective, to include green spaces, entertainment zones, parks, benches and others in their offerings.

For the interviewee, people are not seeking complexity, but simplicity. It is the most basic that has the most value in their eyes. Construction companies should, as a result, widen their scope of how the city of tomorrow should be built and look like, otherwise they may fail to deliver and eventually be evicted from the market. Customers care about the little details and appreciate eco-friendly construction practices. Therefore, everything that generates less pollution and allows for energy consumption reduction is often admired by customers. Besides, construction companies should not pretend to know what people actually want, because they do not. Customers should be part of the process, and it is only when companies and communities cooperate together that large construction projects are destined to succeed. At the building level, the interviewee affirmed, it is important for companies to keep a close eye on the design and modules of the ground floor. In his viewpoint, it is this part of the building that will eventually decide whether the latter would sell or not – whether a neighborhood is livable or not – whether a city is appealing or not. The ground floors of buildings indicate the dynamics of a city. Smart cities, once built, are likely to be job creators; indeed, a community or a property manager must be hired to ensure that all components of the city are interrelated and all systems are working as intended (e.g. waste collection, grid management, energy consumption, space management and others).

Technically speaking, whenever an idea of a large project springs to mind, the execution of that project follows a sequential process that extends from the day the company meets with the public landlord and makes them a proposition about the envisioned project, to getting state permits and permissions, to finally settling through concessions with local communities on a clear expose of what the city would entail and resemble.

In our envisioned BM, integrating the GC as a construction maven in the value chain does not mean that they would have the exclusiveness of fulfilling all works required in the project. Their role is mainly managerial, to oversee and ensure the provision of all necessary trades for the accomplishment of the project. Under the GC BM, as under any other BM, groundworks may possibly be outsourced to third parties who would operate under the supervision of the GC. Also, we note that, nowadays, in France, the world smartness, often misplaced, is being used by construction companies as a mere marketing tool that allows them to sell their final products faster and at higher prices. Until construction companies are held accountable or take responsibility for their actions, the interviewee declared, companies will continue labeling and selling their products as being eco-friendly and smart, while in reality they are not. Corporate social responsibility is a key topic that should be seriously considered in the realm of smart cities, especially for construction companies. Also, given that decision-making is often decentralized in the construction industry, the likelihood of mix-ups occurring in conception and execution plans is high. The interviewee said:

> This gives further credibility to your proposition, as once decision-making is centralized in the hands of one entity, such flaws would be dodged. The maladministration of documentations, often resulting from records dispersed in the hands of numerous stakeholders, would also be escaped.

We assert that under the GC BM, the GC – along with other stakeholders, among them end-users – would take part in both the conception and execution of the project. Smart cities cannot be built without team effort, collaboratively rather than individually. Hence, it is a process of value creation (for all) through value networking (by all). Whenever we work individually (without supervision), each one of us would seek our own interest, yet under the GC BM, the odds of the collective interest being served are higher. This is crucial as most propositions made to customers today lack value. Smart cities are never about the technology nor the structure, but about the humane and the livable. Smart cities are mainly a question of smart governance and smart people. All other components come second. They are an assortment of different things, including the environment. The latter should be given further attention by construction companies. By using both smart technologies or systems and sustainable construction materials, the yield that we would expect at the end of the process would inevitably be eco-friendly, sustainable and smart. And so, smart cities are complete packages, including different components that should be considered and thought of before the project begins. Under our GC BM, GCs could be defined as the entities who are in charge of bringing all city components together, which aligns with the experience of Copenhagen, Denmark, in the domain of smart cities.

The GC BM, we found, is a necessity today in the construction market in France, since most of the companies operating in the market are small and specialized in a particular field. The problems listed thus far seem to be common to almost all construction companies operating in France. The innovation we are suggesting is correctional rather than disruptive. And, we presume, it would be beneficial for all: for project owners because they would have to sign one agreement with only one entity, instead of several, who would handle of all the risks associated with the project; also, for small-sized construction companies who often handle micro-scale construction projects, having only one correspondent to talk to throughout the entire construction process is regarded as an invaluable innovation. And so, we conclude, if such an innovation is required for micro-scale projects, what would it be for macro-scale projects? To conclude my investigation regarding smart cities, I held several meetings with Mr. C.R., a professor at a very well-known University in

Paris, where we both agreed on the fact that the right departure point for any future research on smart cities would consist of deciphering the notion of smart cities, first by embarking upon all academic research on the topic, then by attempting to make an innovative and scientifically proven proposition in this field. Within this context, I, together with Mr. C.R. and a few other scholars, proceeded by creating a committee of researchers (including Ph.D. students and university professors) to conduct further research on the question of smart cities and smart territories, a work that, we hope, would eventually lead to the creation of a research center (Symbioter – as we thought of naming it) that would be specialized in the study of the circular economy, territorial development and human-centric smart cities.

Table 4.4 shows the main problems identified in phase 3 and the solutions suggested for their resolution. Accordingly, a third and final version of the BM has been envisioned – illustrated in Figure 4.4. We reiterate, color codes were meticulously chosen to reflect potential dependencies between the BM's six sequentially ordered constructs: construction stage, BIM process, construction actors, TLBM layers, smart city components and TLBMC building blocks. As you would notice, a technical layer has been added to the TLBMC. It was added to fit in all technical aspects of construction projects, and also to account for all aspects that could not fit in other layers. Besides, the BM's boundaries were colored based on their prospective links with both smart city components and TLBMC layers: financing revolves around the economic, social acceptability around the social, and construction standards and codes – and legal and institutional requirements around the technical. Governance with smart city components has a unique color, as governance could have different forms: social, economic, technical and others.

<table>
<tr><th>Identified problems</th><th>Suggested solutions</th></tr>
<tr><td>– Governments' reticence to openly share data</td><td>External (uncontrollable) factors</td></tr>
<tr><td>– Lack of smart city initiatives and urbanization efforts</td><td>– Encourage smart and eco-friendly initiatives (urban-village)</td></tr>
<tr><td>– Non-sustainable construction</td><td>– Use of innovative, eco-friendly, cost-effective construction materials
– Use additive or other technologies and systems to boost energy efficiency</td></tr>
<tr><td>– Broad concerns apropos the livability of assets</td><td rowspan="2">– Build eco-friendly living environment
– Enhance end-users' wellbeing and quality of life
– Ensure final product has a use-value and that utility may be extracted from it</td></tr>
<tr><td>– CSR-related topics occasionally conversed for mere marketing purposes</td></tr>
</table>

Table 4.4. *Problems versus solutions (phase 3) (source: created by the author)*

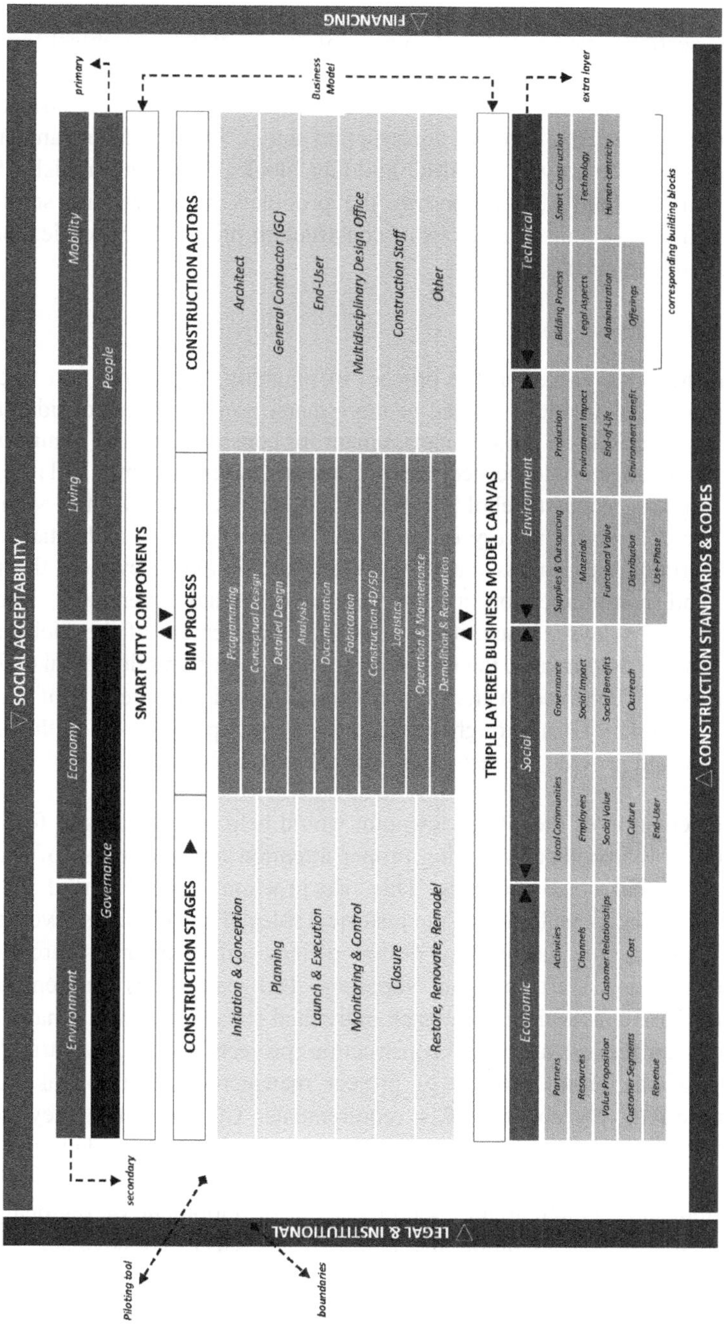

Figure 4.4. *Version 3 of GC BM (source: created by the author). For a color version of this figure, see www.iste.co.uk/karam/general.zip*

4.3.1. *The six-stratum business model*

In the next sections, we briefly explain each of the six strata (variables) forming our envisioned GC BM for the building of smart cities. Following a sequential order, we first embark upon the model's pivotal (first) stratum, which is the construction stage, going through the BIM process, construction actors, TLBMC layers and smart city components, all the way to TLBMC building blocks, being the model's final stratum. The latter would help us portray the problems identified and solutions suggested earlier across four dimensions of construction projects: economic, social, environmental and technical.

4.3.1.1. *Construction stages*

Every stakeholder involved in the process of planning, designing, constructing and operating physical facilities related to the (smart) project under consideration, gains different viewpoints on project management for construction. The contribution of proficient knowledge could be very helpful, mainly when it comes to large and complicated projects, since stakeholders in various specialties could offer valuable services. On the contrary, it is very essential and advantageous to understand how the different parts of the process match together. As previously evoked, the poor coordination and communication among stakeholders may eventually lead to excessive costs and delays. Hence, it is chiefly the requirement of the (delegated) project owner that is, in our case, the GC, to assure that such flaws do not take place. And all stakeholders involved in the project should regard the interests of project owners – represented by GCs throughout the entire project process – as at the end – it is them who would provide the resources and make the decisions.

The implementation of the GC's viewpoint would help stakeholders to focus on the completion of the project by having proper attention on the process of project management for constructed facilities. This, we proclaim, would curtail the old concept of making decisions based on the historical roles of stakeholders involved in the project, namely: multidisciplinary design offices, urban planners, architects, engineers, construction staff, material suppliers, ICT companies and end-users (among others). There is no doubt that each individual stakeholder would have a lot of contribution to the advances of construction projects[8] – yet, we trust, the understanding of the entire process of project management would make them respond more effectively to the GC's requirements. Consequently, they could contribute their proficiency through opinions in improving the productivity and quality of their work. The enhancement of project management boosts the construction industry, which in turn facilitates the development of regional and national economy (thus, cultivating the allure of the city). To have significant

8 Construction Project Life Cycle – Phases in Life Cycle of a Construction Project. Available at:https://theconstructor.org/construction/construction-project-life-cycle-phases/14283/.

improvements, we should be familiar with the construction industry, its working environment and the legal and institutional constraints that are likely to influence its activities and the nature of project management.

Generally, a standard construction project is divided into six major lifecycle phases (see Chapter 2 for more details in this regard) – enumerated and described below:

4.3.1.1.1. Initiation and conception

With any construction project, the first step should be to determine whether or not it is actually feasible and therefore, whether it should go ahead. This is the point at which the project objectives or needs are acknowledged; this could be a business problem or opportunity. A suitable response to the needs is documented in a business case with recommended solution options. Some preliminary feasibility studies are then conducted to examine whether each option clearly addresses the project objectives, and a final recommended solution is then put forward.

Many questions related to issues of feasibility are raised and addressed during this phase, for instance: could and should we do the project? When a solution is approved, a project is initiated to implement the accepted solution. At this stage, the main deliverables and stakeholders that would partake in the process are identified. This is the time when the project team begins to take shape. Approval from the GC is then required in order to move onto the detailed planning phase. GCs, architects, ICT companies and end-users all play a key role during this phase of the construction process.

4.3.1.1.2. Planning

The planning phase involves further detailed development of the project to meet project objectives. The GC identifies all of the work that needs to be achieved. The tasks and resources required for the project are determined, along with the strategy for producing them. In a broader sense, the identification of each activity and its associated resource allocation are also carried out. A project plan outlining the activities, tasks, dependencies and timeframes is created.

The GC – along with other key stakeholders, such as architects, engineers and ICT companies – is the person who handles scope management. This involves coordinating the preparation of a project budget by providing cost estimates for labor, equipment and materials. The estimated budget is then used to monitor and control expenditure during project implementation. Finally, the GC prepares several documents that outline the quality plan, providing quality targets, assurance and control measures – as well as the acceptance plan, listing the criteria to be met to gain customer acceptance. At this point, the project will have been planned in detail and will be ready for launch and execution.

4.3.1.1.3. Launch and execution

During this stage, the project plan is set in motion and work begins on site. The GC spends a considerable amount of time dealing with this part of the process. Throughout the project implementation, different stakeholders carry out a variety of tasks, and progress is reported via regular project team meetings.

4.3.1.1.4. Monitoring and control

The GC would use all information identified in the previous stage to preserve control over the direction of the project, by comparing the progress reports with the project plan to measure the performance of the project activities. If any deviation is noticed from the already defined plan, corrective measures are made. The first option of action should always be to bring the project back to the original plan. Nevertheless, if this is impossible to do, the GC may proceed by recording deviations from the original plan, and recording and publishing adjustments made to the plan. Throughout this step, investors (sponsors) and other key stakeholders (including the original project owner) are kept informed about the project's status. Generally, status reports must always highlight the probable end point in terms of cost, schedule and quality of deliverables. Also, each project deliverable produced should be reviewed for quality and measured against the acceptance criteria. When deliverables have been produced and the customer has agreed on the final solution, the project is said to be ready for closure.

4.3.1.1.5. Closure

During closure, the importance is on providing the final deliverables to end-users, that is, handing over project documentation to the project organization, terminating supplier contracts, releasing project resources, communicating the closure of the project to all stakeholders and conducting lessons-learned studies.

4.3.1.1.6. Restore, renovate, remodel

A few years after closure, there is a chance that the project delivered might require maintenance, thus it is up to the GC (or assigned specialized entity) to decide whether the property should be restored, renovated or remodeled. In a nutshell, restoration is the process in which the building would be returned to its original condition. It includes simple repairs to be made to the property. Renovation, on the contrary, is the process of renewing a building or structure by fixing what is already present and in some cases adding new components. Often, renovations are done in conjunction with restorations, as we could decide to restore a given property and modernize it at the same time. Finally, remodeling involves changing the entire structure of a given property or structure. Relative to restoration and renovation, remodeling is much more labor-intensive and costly.

4.3.1.2. *BIM process*

The BIM process founds stratum 2 of our GC BM. It is directly linked to stratum 1: the construction stage. By combining the two, we allowed for an additional breakdown of construction stages, and so, a better trailing of construction works carried out at each stage. The tie between the two strata is shown in Figure 4.5.

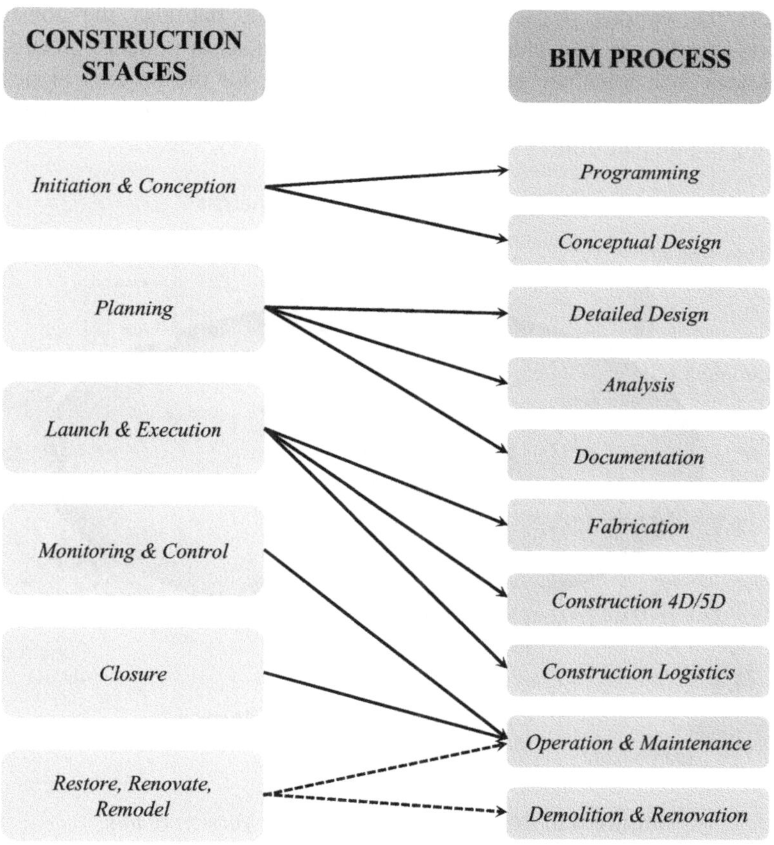

Figure 4.5. *Association between construction stage and BIM process (source: created by the author). For a color version of this figure, see www.iste.co.uk/karam/general.zip*

Given the broadness and immensity of the BIM process (a total of ten phases), it is essential to locate oneself within it[9]. Today, France-based construction companies, mainly those using BIM to keep track of their construction project processes, do not develop BIM building models, but rather BIM models that define and specify the building or structure to be built. Precisely, they do not model the construction site and do not consider the details and phases of the building systems that would be used to build the building or structure. In other words, they model the structure to be built, not the process of building that structure. Hence, we affirm, BIM is often misconstrued and its potential not fully exploited by construction companies. This is factual since the software could be used to develop a BIM model that not only considers the various stages of a construction site, but also the construction processes (see Figure 4.6). Hence, we avow, BIM is a set of policies, processes and technologies that, together, generate a methodology for the process of designing smart constructions, testing their performance, and managing their information and data using digital platforms.

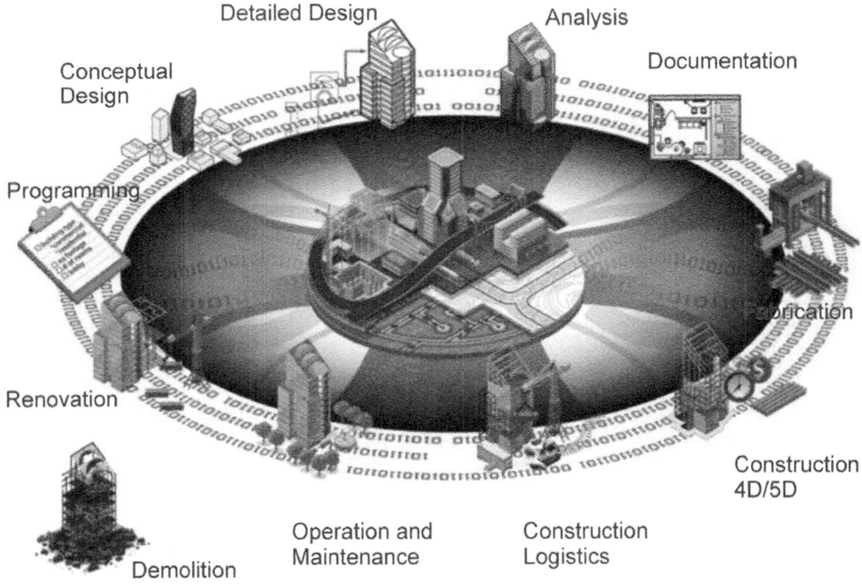

Figure 4.6. *BIM process[10]. For a color version of this figure, see www.iste.co.uk/karam/general.zip*

9 What is BIM? Available at: https://bimmda.com/en/what-is-bim [Accessed June 21, 2021].

10 Building Information Modeling Process. Available at: https://www.realityimt.com/as-built-bim/ [Accessed June 21, 2021].

Besides, BIM is applicable to the entire lifecycle of a construction project, from the conception and creation of a construction project idea, through project development and construction; until after the work is completed, and the final product is delivered and occupied – the reason why we chose to integrate it into our GC BM. Also, based on the scope and objectives of the construction project, BIM may possibly be used to build various models at different stages of a construction project's lifecycle, while considering the consolidation of data resulting from the evolution of the project and the process of defining constructive solutions and specifications. BIM models, we add, may even be used for occupancy management and maintenance. And so, in conclusion, we trust that by integrating BIM into our GC BM, we would be disposed to resourcefully manage smart developments, by avoiding maladministration of project documentation, reducing paperwork, keeping track of every detail throughout the entire lifecycle of a construction project, curbing needless operational costs, and helping to ensure the accurate synchronization and cueing of tasks to be fulfilled by stakeholders involved in the project.

4.3.1.3. *Construction actors*

In construction, numerous stakeholders (or construction actors, the third stratum of our GC BM) join forces throughout the lifecycle of a construction project to build the intended structure and deliver it on time and within the set budget. Aside from the customary roles assumed by architects, engineers, construction staff, multidisciplinary design offices and others – in our GC BM, we instead focus on the role of a GC for the building of smart cities. A GC is an entity or a person hired by the project owner (client) to carry out the work that is required for the completion of a project. Nevertheless, GCs do not always have the skills or expertise needed to complete every aspect of the construction work. Thus, time and again, they appoint subcontractors to carry out particular parts of the project. It goes without saying that the role of GCs has many different levels, as they are responsible for a wide range of tasks and people. Finding the right people to get the job done, and taking care of materials, equipment and any other services required for the smooth running of the project is all part of their remit. This is also where subcontractors can provide valuable help as, in most cases, GCs have their own network of subcontractors who can help to progress the project, and allow for its delivery, possibly within shorter timeframes.

The most common project aspects that a GC could be responsible for include, but are not limited to: (1) applying for building permits; (2) ensuring property security; (3) providing temporary facilities on site; (4) taking care of generated waste; (5) managing on-site personnel, which includes the implementation of safety and security measures; (6) site surveying; (7) site engineering; (8) monitoring schedules and progress; (9) modifying and updating the project program based on most recent changes and feedback; and (10) collaborating with contractors and subcontractors (see Table 4.5).

Manage daily operations	A GC's main job is to manage the day-to-day operations at the construction site. They should also focus on keeping all tradespeople on-task and the work site safe and clean.
Provide all construction materials and equipment	A GC is responsible for providing all of the construction materials and equipment needed for the project. They ensure that the best quality materials are used, while keeping the client's budget in mind. They also make sure that all materials show up on time and in great condition.
Coordinate contractors, subcontractors, and architects	A GC hires specialized subcontractors to work on portions of the project (e.g. plumbers, painters, and others). On the work site, they are responsible for coordinating all of the subcontractors to keep them on task. They also relay important information from the project owner to the subcontractors. Additionally, they are responsible for the timeline, budget, and quality of work — and make weekly check-ins to make sure that all aspects of the project are running smoothly during the entire duration of the build.
Main contact to the client	A GC acts as the primary contact to the project owner, ensuring the latter is kept updated with the progress of the project. From the initial consultation and onward, a GC works with the project owner to guide the project to completion.
Collaboration across the supply chain	A GC works on improving communication and collaboration among stakeholders involved in the construction process which, we trust, could make the whole difference between the success and failure of a project.
Managing the team on site	A GC uses digital tools such as BIM to establish a strong connection between the site and the office. BIM is quite useful as it enables GCs to competently handle and manage several projects at a time.
Task delegation	A GC uses interactive systems to establish connections between office and site, delegate tasks and keep all stakeholders on the same wavelength.
Documentation	Documentation is one of the biggest challenges for every PM. The amount of captured data in the course of a construction project is huge and with this in mind, it becomes clear that the transition to digital is of paramount importance today. Hence, the invaluable role of our *stratum 2* (*—BIM Process*) in this respect.
Learning from past projects	A GC efficiently handles data collected throughout the lifecycle of a project and uses it as a basis for improving efficiency and reducing rework rates in future projects.

Table 4.5. *General Contractor's main duties (source: created by the author)*

Based on the aforesaid, it seems obvious that a GC would also be accountable for the quality of the works delivered to end-users. Under the GC BM, the GC is a direct contractor, an entity who is hired directly by the project owner. In such a case, the relationship of a project owner with a GC takes the form of a long-term commitment, which may sometimes include one or more future projects. Under the GC BM, GCs are also entrusted to help architects plan and design the project, ensure that all work done in a project complies with the construction standards and codes (environment related, social related, safety and health related, etc.), and stay on top of the work done on site. Also, GCs are expected to act like design and build contractors who would be involved in almost every step of the project, and whose job may extend to post-delivery phase, even after the project has been completed and delivered, by, for instance, ensuring the maintenance of the structure built. Hence, under the GC BM, GCs are consigned to concurrently act as CMs and PMs.

Summing up, it is evident that providing a specific definition for the role and the responsibilities of GCs could be harder than it might, look due to the multi-purposed services that they provide during the course of a construction project. However, we proclaim, the advent of technology in combination with the various challenges that the French construction industry is currently facing has paved the way for a new approach in the roles and responsibilities of GCs. Finally, a GC job description may alter on a project basis – based on whether they are assigned to just provide management for a project or also participate in construction.

4.3.1.4. *TLBMC layers*

As discussed in the preceding chapters, the TLBMC is a new strategic management tool that was developed by Pigneur *et al.* (2015) with the aim of exploring sustainability-centered BMI. The TLBMC, built upon the foundation of Osterwalder (2004), adds two new layers to the traditional BMC: environmental and social. This new BMC, we proclaim, moves a company away from the one-dimensional perspective of the old-style BMC by incorporating the three-legged approach, allowing for a more holistic BM to be achieved.

In our GC BM, however, we have added a fourth (technical) layer to the new BMC, thus making it four-legged (stratum 4 of our GC BM). Overall, the new version of the TLBMC enabled us to map, understand, evaluate, design, implement, operate and control four key layers instead of three (and 34 building business blocks instead of 27). The problems identified during the course of the BM design process are allotted across the four TLBMC layers, so we could see within which layers most problems lay, and, consequently, suggest plausible solutions for their resolve.

On a separate note, the innovation we are trying to bring to the construction industry's value chain suits Hart and Milstein's (2003) greening concept, which in the authors' words, is a process whereby a company attempts to harness sustainability through incremental rather than radical innovation. Within this context, we make it clear that the TLBMC allowed us to analyze and conceptualize the dynamics and coherence of our BMI; first, vertically, by comprehending how value is created across all three-plus-one layers, and then, horizontally, by exploring and understanding new ways of creating economic, environmental, social and technical value – individually – via the interaction of the building blocks within each layer. Finally, we say, the TLBMC permitted us to make a significant first step forward towards transforming the old construction industry's BM into a sustainable one.

4.3.1.5. *Smart city components*

Smart city is a new concept that is based on and enabled by technology. As previously evoked, the rapid increase in population and urbanization worldwide have increased the basic ways to manage urbanization, with maximum impact on the environment and lifestyle of inhabitants. The prior integration of new ICT into urban operations has promoted the concepts of digital cities. And the design of the IoT for smart cities resulted in ensuring smart cities' goals with minimum human interactions. With technology playing an increased role in the building of smart cities, cities eventually converted into technology demonstrators and lost their social appeal. To settle on what a smart city really means is not easy, since there is no universally accepted definition of it.

To my knowledge and based on the extensive readings I did on the topic, I can openly affirm that a smart city is expected to mean different things to different people. The conceptualization of a smart city, I add, is likely to change from city to city and country to country, depending on the level of development, willingness to change and reform, resources and aspirations of the city residents. Indeed, a smart city would have a different connotation in India than, say, Europe. Some definitional boundaries are required to guide cities in this mission. This is true, as in the imagination of a smart city, the wish list of infrastructure and services required varies between cities and nations. Nowadays, in an attempt to provide for the aspirations and needs of the people, urban planners are seeking to develop entire urban eco-systems, which are represented by the four pillars of comprehensive development-institutional, physical, social and economic infrastructure. This could be a long-term goal and cities could work towards developing such comprehensive infrastructure, incrementally, by adding some layers of smartness to their existing BMs (see Chapter 1 for more details regarding smart cities).

For the purpose of this research, we assume two primary components of smart cities: governance and people, and four secondary ones: economy, environment, livability and mobility (altogether forming stratum 5 of our GC BM). Some classic features of comprehensive development in smart cities are listed as follows[11]:

1) Housing and inclusiveness (i.e. expand housing opportunities for all).

2) Create walkable localities (e.g. reduce congestion, air pollution and resource depletion, boost local economy, romote interactions and ensure security).

3) Preserve and develop open spaces (e.g. parks, playgrounds and recreational spaces in order to enhance the quality of life of citizens and others).

4) Promote a variety of transport options (e.g. public transport and last mile para-transport connectivity).

5) Make governance citizen-friendly and cost-effective (e.g. rely on online services to bring about accountability and transparency, and form e-groups to listen to people and obtain feedback).

6) Apply smart solutions to infrastructure and services in area-based development, in order to make them better and more resilient (e.g. make areas less vulnerable to disasters, use fewer resources, and provide cheaper services).

4.3.1.6. *TLBMC building blocks*

The TLBMC building blocks constitute the sixth and last stratum of our GC BM. Overall, after adding the technical layer to the BMC, the total number of building blocks rose from 27 to 34, hence allowing for a larger breakdown of problems identified and solutions suggested. The economic layer of the BMC accounts for nine key building blocks causally interacting with one another. At this level, the focus is on the economic sphere, that is, the economic profitability of construction projects. The environmental layer, on the contrary, focuses on the lifecycle assessment, measuring the environmental impact of a structure throughout all stages of its life. At this point, companies should concentrate on identifying material environmental issues through tools, such as energy and water use, waste generation, resource depletion, ecosystems destruction, and negative effects on public health and life. Once a critical material environmental issue has been identified, construction companies should come up with solutions aimed at resolving it. On the social front, the focus is put on the construction company's social impact. By incorporating a stakeholder perspective, construction companies would be better suited to balance the different interests of their key stakeholders. This layer, we avow, is broad and flexible in use. Finally, the newly added technical layer accounts for the technicalities of projects, mainly technical issues faced which, if unsolved, may possibly have some serious repercussions on the other layers and buildings blocks of the BMC.

---

11 Smart City Features. Available at: http://smartcities.gov.in/content/innerpage/smart-city-features.php [Accessed May 8, 2020].

ECONOMIC LAYER				
Partners — an entity (person or company) with which another entity has some form of a business alliance, temporary or perpetual	**Activities** — the action of making, providing, purchasing, or selling products or services	**Value Proposition** — an offering, innovation, service, or feature intended to make a company or product attractive to customers	**Customer Relationship** — the development of an ongoing connection between a company and its customers	**Customer Segments** — the process of dividing customers into groups based on common traits so that companies could market to each group effectively and aptly
	Resources (Human and Nonhuman) — the factors used in producing products or providing services		**Channels** — a chain of businesses or intermediaries that a product or service passes through until it reaches the final customer	
Costs — the amount of money to be spent to obtain or produce something			**Revenues** — the income generated from normal business operations or activities	
ENVIRONMENT LAYER				
Supplies & Outsourcing — the use of the services of an external specialized supplier to complete a given task and achieve green goals as a result. It could also relate to the provision of sustainable products or services.	**Production** — is concerned with the (sustainable) process of transforming a range of inputs into those outputs that are required by the market	**Functional Value** — the solution an offer provides to the customer	**End-of-life** — it denotes the end of a product's useful life; it is when the producer decides to stop marketing, selling, or rework sustaining it	**Use Phase** — The impact that may ensue throughout a product lifecycle, principally when the customer is using the purchased product or service. Recently, the expression 'UX, User Experience' has become a trend
	Materials — the various resources, sustainable or not, used in construction or production processes		**Distribution** — it is concerned with the strategies to be adopted by a company to sustain the delivery process of outputs to markets (finding best ranges, using eco tires, etc.)	
Environmental Impacts — any change to the environment, adverse or beneficial, resulting from a company's activities			**Environmental Benefits** —generating power using renewable energy technologies is an example of an environmental benefit leading to, say: improved air quality, reduced noise pollution and reduced consumption of fossil fuel energy sources	
SOCIAL LAYER				
Local Communities — those people living in a given locality irrespective of their origin and may have some common interests in the area	**Governance** — the set of socially conscious standards on which a company manages its rapports with staff, customers, partners, and the communities where it operates	**Social Value** — the quantification of the relative importance that people place on the changes they experience in their lives	**Societal Culture** — the commonly held beliefs or values that exist and are agreed upon in a given population	**End-User** – the final customer, consumer or client. The term end-user is normally used in IT to denote the person that a software program or hardware device is designed for
	Employees — a person working for another person or a company for pay		**Scale of Outreach** — capacity of a company to reach out to people to promote its propositions through word-of-mouth	
Social Impacts — how companies or individuals' actions affect surrounding communities			**Social Benefits** — literally, they denote transfers received by households intended to provide for the needs that arise from certain events or circumstances. In construction, it concerns improved living standards, greener neighborhoods, etc.	

Table 4.6. *Definition of the TLBMC building blocks (source: created by the author). For a color version of this figure, see www.iste.co.uk/karam/general.zip*

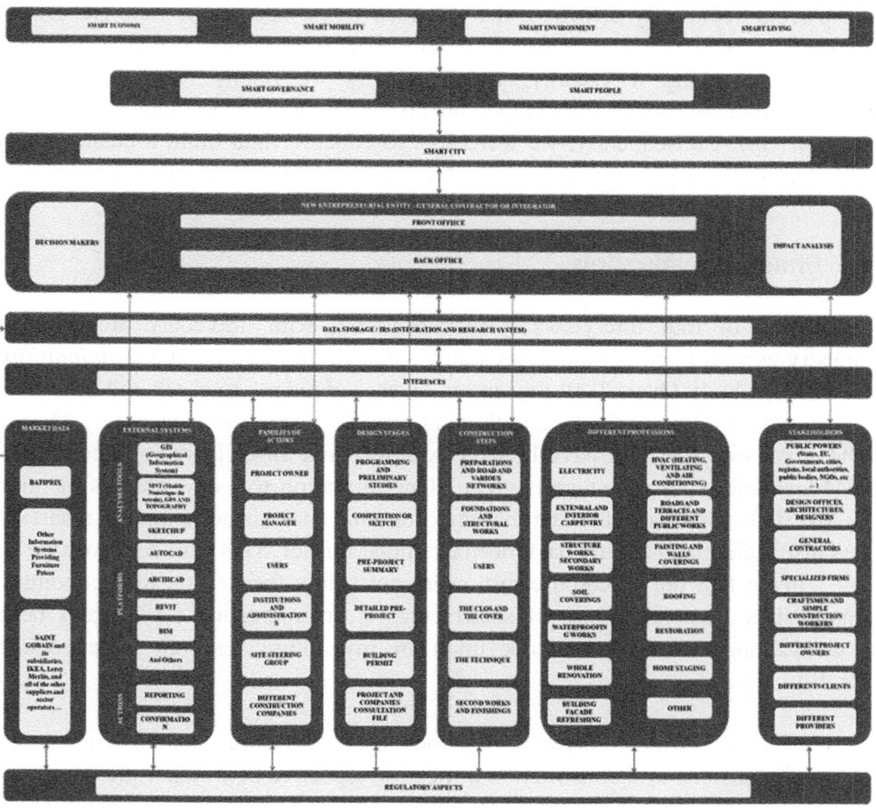

Figure 4.7. *An early version of the envisioned GC BM (source: created by the author). For a color version of this figure, see www.iste.co.uk/karam/general.zip*

For the successful achievement of our empirical research, we sought to identify, map and evaluate all economic, social, environmental and technical issues encountered during the building process of smart cities in general, and constructions in particular. Subsequently, once issues have been identified and mapped out, we used the TLBMC to explore new ways (solutions) to transform the BM through innovations aiming at creating, delivering and capturing superior value.

Before discussing the fourth phase of the design process (evaluation), I conclude the present section by affirming that the idea I first had of the new GC BM was somewhat complex (see Figure 4.7). As a matter of fact, I faced some serious troubles portraying it visually, which is why we finally resorted to using a specialized IT developer, who helped me arrange both my ideas about, and vision of the new BM. Consequently, we were able to conceive and build a business model simulator following the design process phases listed and described earlier.

4.4. Phase 4: evaluation

A total of four interviews were carried out with university professors and industry experts to gather their viewpoints vis-à-vis the *validity* and *sustainability* of the GC BM built. Overall, all four interviewees *valued* my research and avowed that the suggested BMI is original and thought provoking. Precisely, as put by Mr. B.G., "the smart city BM you have built is extensive and all-inclusive – accounting for BIM technology, too – which is great". One limitation was nevertheless mentioned by Mr. B.G., relating to the fact that the economic analysis pertaining to smart cities was insufficient and should have been extended further. This is understandable – we say – given that our research efforts were not solely focused on the economics of smart cities. The economics of smart cities is an interesting topic that I believe I would unquestionably embark upon in my future research endeavors. For Mr. B.G.:

> The GC BM is inventive in the sense that it accounted for all dimensions of a construction project, irrespective of whether the latter is smart or not. The integration of the GC role as a central operator within the construction industry's value chain is ahead of the functioning of our organizations, who are still a bit orthodox in their ways of thinking.

Also, he admired the business model simulator (software scripted in Agular 2+), which, in his opinion, could be used along with other available technologies to ensure the resourceful planning, execution and management of future smart city projects.

> This study is a robust starting point that France based construction companies may eventually rely on to improve their construction practices and contribute to the emergence of a real smart city market in the country.

In the same manner, Mr. J.Y. avowed that the suggested GC BM is very interesting, especially as – in France, it is uncommon to find someone who is capable of looking at things from a different angle. He said: "though the GC concept is practical and viable, the risk market players would refute it is high; the market itself does not allow for such innovations – which is unfortunate".

Even though the GC concept exists (it actually does exist!), he would be forced to outsource construction works – and so his job would be limited to control and management.

> The fact that you added the BIM into your BM is a smart move. Also, the ability of the business model simulator to generate changed BMs, showing possible interactions among stakeholders, is remarkable as problems and solutions could then be mapped stage by stage.

He finally suggested testing the GC BM in real-life to see how it works and whether or not further improvements should be made to it. Similarly, Mr. P.P. affirmed that he has never seen such a tool before (referring to the business model simulator), insisting that the suggested GC BM brings some real and useful changes to the industry's old BM. He said: "the construction sector is never focused on improving its ways of doing, which partly explains why we are still committing the same mistakes we used to do fifty years ago".

Moving forward, Mr. C.R. avowed that the research I carried out is perfectly structured, including my choice of the business-modeling tool. In his eyes,

> The TLMBC is a somewhat perfect tool that one could use to map problems faced at industry level today and suggest solutions for their resolution. By adding end-users as construction actors, you have respected the human-centricity approach to building smart cities where people partake in the process as both data sources and innovation creators.

> The TLBMC *per se* responds to all aspects of a construction project, smart or not, and the technical layer, I admit, is a beautiful addition to your BM or, better named, piloting tool.

In Chapter 5, we discuss and map the main problems in construction, as well as the solutions suggested for their resolution. We also clarify and graphically illustrate the existing dependencies among the BM's layers and between its building blocks.

5

Research Findings

This chapter identifies the main problems encountered in construction today and suggests plausible solutions to resolve them.

This chapter includes contributions from various industry experts and scholars who the author interviewed using the IDI technique. The identities of the interviewees have been kept confidential throughout the whole text, and so we refer to them by their respective initials rather than full names. Further information on this and the IDI technique can be found in the appendix at the end of this chapter (section 5.5).

5.1. Problems

We herein present an inventory of the main problems encountered by construction companies today. We use the TLBMC's four layers – *economic*, *environment*, *social*, and *technical* – to map and explain them. As you can see throughout the text, the problems identified are also split into two categories: *controllable* and *uncontrollable*. In our views, *controllable problems* simply denote project management problems that could be addressed and solved by construction companies once the appropriate measures have been implemented. However, *uncontrollable ones* – which we label O-B (Off-Balance) – are those that construction companies have zero control over, yet that could markedly affect their capacities planning in project management and put their project management structures in question. Urbanization, financing, legal and institutional barriers and state-imposed construction codes and standards are examples of uncontrollable, off-balance problems. This distinction between controllable and uncontrollable problems is vital in the sense that it allows construction companies to separate operational problems from institutional and other macro ones. Note that problems in construction repeatedly ensue from institutional problems – being out of construction companies' control. Over time, those problems, we believe, are undoubtedly going to magnify and create other, operational, problems such as budget overruns and delays in deliveries.

In the following sections, we group the various problems identified under specific headers: economic, social, environmental and technical – and explain them thoroughly.

5.1.1. *Economic*

We herein list and explain the economic problems faced in construction. First, we tackle off-balance or uncontrollable problems – and then embark upon controllable ones.

5.1.1.1. *Financing (O-B)*

Financing plays a vital role in the building of smart cities. David Harvey, a neo-Marxist-inspired author who took a keen interest in the study of urbanization of capital, spoke favorably about this[1]. When it comes to France in particular, the process of founding new cities has moved through diverse phases over time. First, there was the Fordist cycle where the state played a chief role in the implementation of urban policies and regional plans. Then, in the 1970s, public finance and rivalry between territories rose as topics of interest for every person. Subsequently, in the late 1980s came the so-called urban entrepreneurship dogma where rivalry started shifting from territory to city level. French authorities strained year-after-year to institute some developmental balance between the country's different territories – but somehow failed to do so.

They were unsuccessful because their strategy consisted of undressing Paris in an endeavor to dress up other declining cities. Looking at the experience of the United States, the United Kingdom or other countries in this area – we could realize that smart city markets in Anglosphere economies are governed by market-oriented policies, where urban governance is often ensured by means of operative private–public partnerships (which is not quite the case in France!). In France, big operators such as Altarea Cogedim, Unibail-Rodamco and ICade, all being listed in the stock market as real estate investment companies, are expected to drive urban growth at country level in the near future.

> By setting appropriate financing schemes for the national real estate market, we would be able to come up with a novel production system that would help our declining or marginalized cities to revive and flourish (Mr. F.C.).

We need to invest to generate more money, and with more money, we would be able to invest more; it is a circular process where more investments lead to more value

1 The Application of Harvey's Theories into China's Urbanization of Capital. Available at: http://lup.lub.lu.se/luur/download?func=downloadFile&recordOId=8979923&fileOId=8979940 [Accessed June 21, 2021]. Other useful insights available at: http://www.globalconstructionreview.com/.

creation. Today, a lot of construction companies are building residential units, but are not making any successful pitches. Does this mean that their construction processes are unproductive? Not really. More likely, their BMs are running out of gas.

> The act of creating smart cities through the implementation of a revised BM is indeed an innovative idea, however to be able to properly implement it, one key factor (among others) should be always considered: financing (Mr. E.N.).

5.1.1.2. *Failing BM leading to abridged growth prospects at industry level*

Generally, urban developments are not smart unless they match people's changing lifestyles. Whether in Paris or in any other city in France – a smart city should be based on modular artifacts that can be adjusted over time. Without question, European architecture is referenced throughout the world[2] – and the architects and engineers who built our capitals and towns were visionaries. Customarily, architects were able to conceive cities based on laid-back urban plans following the instructions of just one entity, an elected official who decreed the structure of the entire conurbation. We should operate in a comparable way in the present day: first, by stopping selling end-users empty ideas and promises – and second, by reviewing our offerings in an effort to bridge the existing supply–demand gap.

> There is no perfect smart city model as there would always be curbs to what we conceive no matter how smart it is (Mr. V.B.).

Today, what is holding France back from building its own smart cities is the widespread fallacy among industry experts of what a smart city really is. Structure wise, we proclaim, horizontal constructions are far from being green or smart – and in France, it would be more conceivable to opt for vertical constructions[3] whenever a plan is set to kick off the smart city business.

In addition, urban variations such as the ones we could find in China (e.g. Shenzhen, Guangzhou) and the rest of Asia –urban villages[4] – should be endorsed in

2 The Complete Guide to European Architecture. Available at: https://www.rolcruise.co.uk/blog/the-complete-guide-to-european-architecture [Accessed June 21, 2021].

3 Vertical City Concept: How to Live a Sustainable Life. Available at: https://www.smartcitiesdive.com/ex/sustainablecitiescollective/vertical-city-concept-how-live-sustainable-life/1163942/ [Accessed June 21, 2021]. Other relevant publications available at: https://publications.iadb.org/en/road-toward-smart-cities-migrating-traditional-city-management-smart-city.

4 China's mass urbanization projects mean the end for Guangzhou's 800-year-old urban villages. Available at: https://www.scmp.com/economy/china-economy/article/3006303/chinas-mass-urbanisation-projects-mean-end-guangzhous-1000 [Accessed June 21, 2021].

France, as, at the end of the day, smart cities are simply incarnations of the humane and the livable. If we appositely construct a 40-storey building that includes small-sized residential units, that is located in proximity to bus and train stations, and grocery stores, and that grants residents access to green spaces and shared entertainment areas, we would then be on the right track to converting our existing cities into smart ones. Smart cities are first and foremost a question of urban governance that should be addressed more seriously by public authorities. If we were to compare the two cities of Lyon and Marseille, though they are of nearly equal sizes, they do not share the same urban histories, nor the same architecture structure types or economic trajectories. While Lyon may have a well-structured urban governance scheme, this is not quite the case in Marseille. Largely, all French cities are currently vying to boost their lure by attracting direct investments towards their own territories. The Bilbao effect[5], a phenomenon whereby cultural investment plus showy architecture are supposed to equal economic uplift for cities down on their luck, is thought to be the direct (and only) solution to the issues currently being encountered in France (which is not quite accurate!). Nevertheless, to trust that via the implementation of megaprojects, declining cities would be able to recover their positions is a huge mistake that local public authorities often commit.

> What people really want and what local authorities actually assume they want are totally two different things. Assuming equality between the two is a gaffe that should be avoided at all costs (Mr. F.C.).

The inadequacy of prevailing governance schemes used in France for the handling of large construction projects is among the key issues that are holding the industry back from success. Indeed, we say, what works for a specific project does not necessarily work for another dissimilar project. Though this is largely factual, in France, project owners, being public utility companies, often opt for a single managerial approach for the running of all construction projects irrespective of their types and sizes.

5.1.1.3. *Lack of understanding of the smart construction process*

The smart city concept is vague. Often, smart cities give the impression that what is being discussed is something very complicated. In France, as elsewhere in the world, they have come up with ample sets of environmental rules and regulations (e.g. carbon emissions, energy consumption per building, etc.), business ethics guidelines and systems values that could be used to competently govern the operational activities of construction companies and become hypercompetitive in the smart city market. In addition, they have satisfactory technological tools that could

5 The Bilbao Effect: is 'starchitecture' all it is cracked up to be? Available at: https://www.theguardian.com/cities/2015/apr/30/bilbao-effect-gehry-guggenheim-history-cities-50-buildings [Accessed June 21, 2021].

be used to establish a healthier environment, make cohabitation easier, encourage social housing initiatives and make buildings more pleasant dwellings. They pretty much have everything, and know how to do almost everything; they have the required knowledge and technology to build smart cities, yet what they do not have is a clear-cut definition of what a smart city really is and what it entails.

> A city is cognitive, smart, intelligent, etc. – a list of echoing yet empty words (Mr. E. N.).

The city is a blend of two very basic concepts: the people who design it and the people who live in it. Which is best, the city that was built 300 years ago or the one of today? Frankly, no one cares about the construction as such. What is significant is the impact and the grip that remains once the construction works are completed.

> The Romans are the ones who built our cities, and, we, ourselves, did nothing but refurbish them (Mr. E.N.).

A smart city is not actually smart, but rather nice and livable. Once we understand this genuineness and consider it, we would then be able to create a humane smart city and regulate its parameters to the satisfaction and fulfillment of its inhabitants. We also add that people repeatedly confuse wants and needs. Thus, we say, the problem we are encountering today does not solely relate to the ability of construction companies to adapt their offerings to people's real needs since, people, time and again, get attached to stuff they do not really need – but have been told they cannot live without. What people really want is to have more trashcans around the school. With no intricacy involved whatsoever, the smart city notion revolves around such rudimentary themes.

5.1.1.4. *Fragmented projects*

By comparing old Parisian buildings to new ones, we can plainly see that the way construction works are achieved has pointedly altered over time, often to the worse, quality-wise. Today, the interest of public authorities is shifting towards construction projects that can be delivered in relatively shorter time frames. As for long-term smart city projects, their budding effect on the attractiveness of cities and the well-being of citizens is thought to be huge, which justifies public authorities' snowballing interest in them. To render our cities smart does not imply that we have to demolish and rebuild them from the ground up. It is rather a question of creating ties between the various components of a city so as to build a coherent system that people could extract real value from.

> Smartness has to do with cultivating people's quality of life and well-being, and its constructs are those that bring use, functional and social value to the people (Mr. D.B.).

The design of the smart city should be carried out in the early stages of the construction process – while financial obligations should be delayed, settled a year or two after the project is finalized. Moreover, in a smart city project, we should discern how to integrate externalities and set an optimal cost-sharing scheme among all the stakeholders involved. When a smart building is set up, the cost of the property must include all the operational costs incurred by the project – or else there would be nothing smart about it. Therefore, construction companies must operate based on a total-cost logic – as a smart city is nothing but a series of ecosystems, a platform that creates wealth, including interlocked components that move harmoniously[6]. Furthermore, the platform built should be neutralizing by nature, meaning that its components must be conceived so they counteract the prospective (negative) effects of one another.

> A smart city is not an architectural model, but a set of criteria that perfectly fit together. Île-de-France (Paris Region) is an example of a region that is smartly governed – but that is not smart *per se*. The key problem we are facing today relates to the fact that most structures in the Paris Region are conceived in silos, not as part of a bigger something. We build, but fail to sell – what is the point then? (Mr. D.T.).

5.1.1.5. *Constricted approach to value creation*

It would be hard to fully grasp the smart city concept – in the French context – unless we have a thorough understanding of how the country's construction market is structured and regulated. If we were to make a comparison between Germany and France – being the two largest economies in the European Union now that the United Kingdom is out of the league, we could straightforwardly assert that, in Germany, real estate represents 20% of a household's monthly budget (on average) – while this figure may go up to 30% in France. Property prices are higher in France because the sector is highly subsidized.

> The more subsidies the sector gets – the higher property prices would be (Mr. V.B.).

In Europe, the more you move towards the South, the more likely it is for residents to be homeowners (e.g. France), whilst in northern countries, residents are most likely to be tenants (e.g. Germany). This is quite reasonable because real estate

---

6 The making of a smart city: best practices across Europe. Available at: https://smartcities-infosystem.eu/sites/default/files/document/the_making_of_a_smart_city_-_best_practices_across_europe.pdf [Accessed May 12, 2021].

is not as profitable in Germany as it is in France. In contrast to Germany, France, we maintain, is an individualist[7] – a southerner – and an owner.

> In France, not only are construction companies blinded by the short term, but customers too – often buying properties for speculative rather than housing purposes (Mr. V.B.).

In Germany, real estate is a common good – a commodity that people do not venture on. In France, however, people operate based on a purely economic logic that often disregards the social and the environmental. The short rather than the long run[8] fascinates them – which, to a certain extent, refutes the basic principles that characterize smart cities.

> Taxing speculative behavior in land and real estate markets may be a practical approach for local authorities to gain a certain control over the situation (Mr. V.B.).

The smart city concept, we finally proclaim, relates to the social as much as it does to the economic, environmental and technical. In fact, smartness is associated with the social well-being of people. Today, construction companies endeavor to enhance people's social well-being through value propositions they believe would meet their needs, yet it is unsurprising that their forecasts end up being off beam – not corresponding to people's real aspirations.

> The smart city notion is highly subjective as different individuals could have different discernments of what it may connote. For developers, smart cities could signify one thing, whereas for end-users, smartness could entail totally different things (Mr. D.B.).

5.1.1.6. *Project process confused for business model*

In France, construction processes are often complicated and unnecessarily arduous. To be more explicit, the current construction value chain is dysfunctional in that value creation is repeatedly forgotten. Hence, the problem is institutional by nature. To say that any loss at any level of the value chain would eventually translate into a loss for the entire chain. This is genuine as the construction value chain is

7 Comparison France–Germany with Hofstede. Available at: http://carnets.parisdescartes.fr/blog/view/325598/comparison-francegermany-with-hofstede [Accessed June 21, 2021].

8 Building the smart cities of the future: Think long-term and local. Available at: https://www.smartcitiesworld.net/opinions/opinions/building-the-smart-cities-of-the-future-think-long-term-and-local [Accessed June 21, 2021].

(or at least should be!) a sequence of interlocked phases where various stakeholders intervene at different levels of the process to fulfill assigned tasks for the delivery of a flawless artifact at the end of the process. Today, the productivity and efficiency of the entire construction value chain are at stake – especially in the knowledge that smart developments are sometimes misconceived and misinterpreted.

> In France, construction actors recognize smart constructions as complex rather than simple projects – though, in reality, most smart city projects, at least the ones we see today and are labeled as such, are very primitive (Mrs. A.T.).

Moving on, there is a tendency today among French-based construction companies to confuse construction processes for business models. The former is intermittent, whereas the latter is perpetual; also, construction processes could vary on a project-by-project basis depending on the specificities of each project, which is not the case for business models.

> Business models are supposed to change every three to five years – based on market circumstances. The life span of a business model is shorter than that of technology – and surely of that of a smart construction project, therefore, whenever crafting a new business model, we must ensure the final artefact is malleable and adjustable (Mrs. J.N.).

5.1.1.7. *Top-down managerial approach*

In France, there are general contractors (GCs) for specific construction branches – mainly in industrial buildings – but, overall, the GC role does not exist in the construction sector *per se*. In the United States and other Anglosphere countries however, the GC role is popular. Largely, in these countries, GCs have some sort of influence over the entire construction process, or even the construction value chain. Specifically, they control budgets and handle the sequence of activities in construction projects, from design, through development and construction, to promotion.

> To introduce such a role into the French construction market is an innovation by itself. At my company, we have been more and more operating based on smart-city-concept logic – working on building smart eco-districts and neighborhoods – thus, for us, such an innovation is concurrently relevant and pertinent (Mr. B.G.).

In our eyes, inept public governance is the main problem causing the issues with the French construction industry today[9]. Though the city of Paris has a powerful standing, it is often administered based on an unbending governance model[10]. For example, based on where we would like to build structures, we might have to deal with three to six different public departments to get the required approvals. Consequently, between the metropolis of Greater Paris and the public authorities concerned, there are over a hundred executives to be approached – thus, we wonder, *how could one ever make everyone agree on a given project*?

> A consensus must be put in place describing how the resources we have must be used for the building of the city of tomorrow. A smart city is a vision of how we, as a society, want our cities to look in the future – and based on this vision, we need to join forces to overcome all barriers that are holding us back from building it (Mr. P.B.).

On the project-management front, managing construction projects is not an easy thing to do[11]. There are many junctures (e.g. design, execution, monitoring, marketing, etc.) that construction projects go through before completion. Not to mention the many state-imposed rules and regulations (relating to the environment, energy production procedures, infrastructural design and others) that construction companies must comply with at all times. Once a company secures a concession contract, there is a lot of work (project diagnostics) to be accomplished before the construction phase – principally feasibility and other preliminary market studies (in cooperation with elected officials and citizens) – in order to meticulously assess the conditions and settings of the construction project.

> Consulting with public authorities, especially local ones, is a critical matter that should be fulfilled before initiating any construction work. This is how we, at our company, conceive and execute complex construction projects – collaboratively, with the assistance of local communities, developers and departments (Mrs. S.P.).

---

9 Smart cities towards a French strategy. Available at: https://inspire.ec.europa.eu/sites/default/files/presentations/377_pdf_copy_of_presentation.pdf [Accessed June 21, 2021].

10 Top-down vs bottom-up planning: Which one is better? Available at: https://www.letsbuild.com/blog/top-down-vs-bottom-up-planning [Accessed June 21, 2021].

11 A solution to handle large and complex construction projects: Interface Management. Available at: https://www.designingbuildings.co.uk/wiki/A_solution_to_handle_large_and_complex_construction_projects:_Interface_Management [Accessed June 21, 2021].

Furthermore, what is specious today is that construction companies tend to make baseless abstractions by imagining that end-users would eventually want from the final product delivered – and that use and intended value will eventually intersect. Buying a house is not the same as buying a car. Generally, a person buys a car without taking part in the manufacturing process of the car (the customer has no say in this respect!). Of the cars available for sale, the customer simply chooses the one that best fits their taste and budget. One seems to trust the oligarchy of companies that manufacture cars. This is not exactly how things work in the construction sector though.

> Buying a property, in comparison to a car, is a lifetime investment; also, the offerings in the construction sector are not as varied as in the automotive sector. Hence, like the automotive sector, the construction sector should allow for further diversification and customization in its offerings so it better succeeds to catch growth opportunities in the market (Mr. V.B.).

Indeed, not all people have the same aspirations and tastes, with some of them finding happiness residing in big cities while others prefer to live in rather quietier and less-populated areas. The main idea here is of a cultural nature, revolving around what people really aim for.

> Human-centricity is a primary component of smart cities (Mr. O.A.).

> In the same manner, social acceptability is a prerequisite for the successful building of smart cities (Mr. P.B.).

Though we have the ability to make our offerings more appealing to people, we are currently lagging behind in a state of imbalance – and to introduce a central entity into the scene that could ensure a fair interplay among stakeholders and enrich the sector's value propositions, we trust, may perhaps be sufficient to help us improve the existing construction value chain.

> Under the GC BM, a bigger role should be reserved for end-users; it is a question of setting up the appropriate steering systems and settling on how information could be centralized and efficiently communicated to customers (Mr. P.B.).

Therefore, what France needs today is a framework by which it could improve communication channels in and outside its cities and inspire value networking by boosting long-term partnerships between stakeholders – including end-users.

> Smart cities could not be built unless all its elements are aggregated together (Mrs. S.P.).

5.1.1.8. *Big companies' resistance to change*

Well-established companies are not necessarily the ones that should handle and manage smart developments as they often take conservative standpoints with respect to innovation. Relative to small-sized companies, large ones are less innovation-friendly. This is true because any change initiative, once applied, can probably upset their stranded moneymaking BMs. Indeed, we proclaim, risk aversion is a barrier to innovation[12]. Moreover, in comparison to small-sized companies, large-sized ones have inflexible governance structures as well as fewer maneuvering abilities to espouse incipient innovations smoothly. This is without mentioning their complex corporate structures which boost their odds of facing exceptional business management problems. Bouygues, we trust, constitutes a great example of a France-based conglomerate – a combination of multiple business entities operating in entirely different industries (e.g. telecom, construction and others) under one corporate group, with each entity having its own management structure, business operations and goals.

> Bouygues is currently facing some serious issues controlling the works of its subsidiaries which partially sheds light on its increasing reliance in recent days on external outsourcing. A GC BM for the building of smart cities is sort of a new-fangled offer of urban services (Mr. B.G.).

5.1.1.9. *Decentralized decision-making (in the management of smart construction projects)*

Today, France is still marching against the stream. Indeed, the country has left the collective world and is now opting for an individualist vision. Put differently, we are showing partiality towards individualism, an ideology that reigned in the 1980s, whereas today, the world instead rather prefers collectivism, referring to collaborative thinking and intelligence.

---

12 Overcoming the aversion to risk key to innovation in construction. Available at: https://canada.constructconnect.com/dcn/news/associations/2018/09/overcoming-aversion-risk-key-innovation-construction [Accessed June 21, 2021].

> To bring change to the existing business model and enhance partnership among stakeholders would undoubtedly do us good. We surely need an entity today that could control and oversee construction works (Mr. V.B.).

Let us add that public developers are often incompetent in France. They often take a leaf out of each other's books and make copies of them – as if all construction projects were the same, involving the same specs – delivered to the same clients. The same applies for engineers who turned into creators of problems rather than of solutions.

> There is a missing link at the level of the construction value chain today. In my view, what is missing is a centralized decision-making model where a maven central operator may be consigned to take the lead and ensure a fair interplay among all construction actors involved in the process (Mr. E.N.).

Today, the existing construction value chain is wrecked in the sense that no continuity exists in the works fulfilled by the various stakeholders, as each one of them tends to operate independently of the others. In addition, knowing that the construction industry is very fragmented and project-based, the odds that value creation would be lost through construction processes are high. Moreover, value networking is restrained because, for each new project commissioned, construction companies tend to bring in new construction staff to handle the execution of on-site works. This frequently translates into broken learning and feedback loops among staff, high turnover rates and diseconomies of scale.

> Public developers, being incompetent in general, opt for external outsourcing for the achievement and management of large construction projects. Usually, large construction works are entrusted to three or more companies rather than one – in French, the so-called '*ensemblier*'. The GC role you are suggesting in the present study would play a somehow similar role to that of an '*ensemblier*' – by synchronizing the works of stakeholders and handling cueing throughout the entire project lifecycle (Mr. A.).

5.1.1.10. *High number of stakeholders with overlapping roles and responsibilities*

On-site construction works are often handled by several stakeholders all seeking to partake in the management of projects. Often, instead of having one decision-maker who could synchronize and competently manage construction works, we often end up with several, each with divergent interests and visions around the project – which is problematic *per se*. The problem does not directly relate to the number of stakeholders involved in the process, but to the fact that stakeholders tend to work disjointedly from one another. Such behavior often translates into some serious operational and financial challenges (e.g. delayed deliveries, wrong procedures, extra costs and others)[13].

> We often see incompetent projects managers, mainly architects, assigned to oversee construction works and manage the project in its entirety. As architects, they do the necessary drawings at the very early stages of the construction process, yet they show lots of inexperience and unskillfulness in all stages that come after (Mrs. S.P.).

Moreover, based on my professional experience, I could easily assert that the larger the number of construction actors involved in the construction process gets, the higher the odds of overlapping roles and responsibilities would be.

> Whenever the number of partakers in the construction process exceeds a certain threshold, every sense of professional responsibility would be lost among them – and diffusion of responsibility would consequently ensue (Mrs. S.P.).

5.1.1.11. *Lack of coordination between construction actors*

Building a smart city is an iterative process where different stakeholders synchronically interact with one another. Lots of excavations and landfills are necessary prior to building large structures. Generally, a construction project is never finished until the moment project owners or whoever is handling the project hand the keys over to the end-users. Smart projects – often long-term constructions – may extend over a 15-year period or even more. Hence, such projects require meticulous consultation and deliberation among stakeholders – including customers and local public authorities –to ensure the envisioned artifact ends up being suitable. Generally, the conception-and-initiation phase starts with an idea and grows to become an iterative process where the first idea shared is repetitively revised and changed. The final framework of the project is expected to take several years being tied up as there are often continual technical and legal makeovers levied by local authorities.

> Construction companies should be aware to consult with local public authorities on the specificities of projects before groundworks kick-off (Mr. V.B.).

---

13 Centrally located: rebranding a decentralized organization with a centralized project structure. Available at: https://brandactive.com/rebranding-project-management-decentralized-organizations/ [Accessed June 21, 2021].

Likewise

> Besides the economic viability of a project, social and political acceptability are two key metrics decisive of its success or failure (Mrs. S.P.).

5.1.1.12. *Loss of control over quality of works*

One of the main issues in the French construction sector today is the lack of human control over construction procedures. To say the least, the smart city concept, at least as perceived in the country, remains distanced from people's real aspirations.

> The output of construction works we are undertaking today will not materialize before a 30-year period. And it is only within this length of time that we would be able to settle whether the structures built have been relevant to people's real aspirations or not (Mr. E.N.).

With several stakeholders stepping in and out of the value chain at different times and stages, it is often hard to trace the quality of works achieved by every intervenor, hence the importance of integrating *quality control* in the management of construction work processes.

> The absence of a central operator who could foresee construction works is indeed a problematic matter in France, thus the importance of the GC role who may possibly fulfill such a task (Mr. D.T.).

5.1.1.13. *Innovation mismanagement*

Construction actors operate either down or upstream of the value chain – not both. Hence, when a construction actor operating downstream makes a promise that the product delivered at the end of the process will have certain value qualities, he often fails to keep his promise, as his contribution at the advanced stages of the process is very limited. At advanced stages of the process, actors operating downstream lose control over the quality of the final product delivered. The same applies to construction actors working upstream of the chain who cannot guarantee the quality of the works performed at the very early stages of the process.

> Today, some big operators such as Bouygues, with the help of their subsidiaries, are seeking to overcome this particular problem by positioning themselves at different stages of the value chain, down to upstream (Mr. B.G.).

5.1.1.14. *Mismatch between promised and delivered final product*

Today, end-users are often dissatisfied with the quality of the final product they get. Often, they end up paying money in return for a low-quality output that is different than the one promised at start.

> End-users are positioned at the end of the value chain and what they eventually get is nearly 90% of what they initially paid for (i.e. the building and the roads), however what they do not get is a sustainable high-quality final product (Mr. V.B.).

Anyhow, we proclaim, developers and subcontractors are not the only ones to be blamed for such defects as end-users should be held accountable, too. What is astonishing is that people frequently choose their properties much faster than they choose their cars. Indeed, home buying is done passively rather than actively, fairly quickly relative to the amount paid.

> When you want to buy a TV, you visit different stores, compare prices, and check reviews to be able to make the right call – yet, surprisingly, this is not how things happen in construction (Mr. V.B.).

5.1.1.15. *Absence of an entity that could orchestrate time horizons and handle maintenance works*[14]

Often, construction companies operate in a purely moneymaking logic. To be more precise, developers often hire low-skilled, economically fragile actors to handle large developments.

> It is a political game that involves the most and least fragile (Mr. V.B.).

After construction has started, if the project is estimated to fail, the developers (being the least fragile) may proceed by canceling it, sometimes without even paying the subcontractors (being the most fragile). Whenever such a scenario concretizes, subcontractors respond by providing poor on-site workmanship. Moreover, in some specific cases, they would even damage the property being built, rendering it unsaleable. If, subsequently, they were paid, they would make the necessary repairs, otherwise end-users would suffer the consequences (e.g. delayed

14 Though this specific problem could have been merged with another one, we chose not to do so in an effort to shed light on the importance of having a specialized entity in large construction projects who could handle follow-up with end-users and maintenance works post-delivery of the final product. We finally note that maintenance works may possibly be handled by the GC themselves, or another specialized subcontractor.

deliveries, damaged final product, and others) as no other construction actor would claim responsibility for such flaws.

5.1.1.16. *Faulty estimates leading to recurrent budget overruns*

Budgeting is another key topic that should be raised whenever attempting to build a smart city[15].

> In France, the idea of budgeting remains very vague, whereas, in Anglosphere economies, it is a present-day topic that occurs at each and every stage of a construction process (Mr. F.C.).

Macro-level budgeting is often misrepresentative, especially in the construction sector, leading to unwanted project budget and time overruns. Alternatively, done properly, micro-level budgeting should start with the tiniest details of a construction project (i.e. negotiations, profit margins, social impacts, environment impacts, etc.). The use of technology in this respect – such as the BIM for instance – we trust to give budget estimates a certain level of accuracy.

> The building of smart cities in France follows a somewhat conservative approach that is driven and controlled by state authorities – which is not quite the case in Anglosphere economies (Mr. F.C.).

Indeed, French-based construction companies are currently operating based on a public business model that is indicted by public authorities – and framed by the texts of public laws. In construction, we specify, there are two types of stakeholders that should be considered: those who invest – and those who do not. Those who invest are those who can contribute money, are business savvy, think in financial terms and seek to minimize costs and maximize profits. However, non-investors are those who contribute labor or technology.

> The GC BM Elie Karam suggests for the building of smart cities would eventually lead to the rejection of the existing industry-level business model since the latter does not account for the main components of smart cities (Mr. P.B.).

The construction of a smart city, we avow, follows a sequential process and is founded on meticulous programming activities. Budget wise, we avow, it is practically impossible for one to predict the exact budget of a new construction project in absolute terms, especially large projects. As a rule, the provisional budget

15 Six strategies to keep your project on budget. Available at: https://www.bdcnetwork.com/blog/6-strategies-keep-your-project-budget [Accessed June 21, 2021].

estimates that PMs or whoever else is handling the project make before the start of construction projects are usually uncertain and erroneous (underestimated). Anyhow, this does not constitute a problem by itself because, in the special case of state-financed projects, there would always be a minority of people who would benefit from such flawed estimates for personal gain. Again, the political comes into play at this level.

> For the Grand Paris Express for instance, though nobody knew the exact budget that would be allotted for the project, the latter was commissioned anyway because it was politically legit (Mr. V.B.).

Putting state-financed projects aside, faulty budget estimates ordinarily lead to a rebuff of construction projects. Indeed, though construction companies usually have a certain maneuvering margin to calibrate their spending so it fits within the budget, such calibrations would not remain possible beyond a given budget-spending limit. The Sydney Opera House is a great example of a chaotic construction project where the budget estimated was just one-tenth of the amount actually spent.

> Within the same framework, it is practically impossible for a construction company not to deviate from the original budget whenever handling a complex construction project (Mrs. J.N.).

Several constraints make it difficult for construction companies to stick to the budget over the lifecycle of a project – one of which relates to the fact that project specs are not always fully detailed and explicitly itemized before the start of the construction phase. Usually, there is an original construction plan and a budget estimate that are settled upon pre-launch of the groundworks. Often, these estimates end up being wrong; hence, adjustments are made to restore the initial figures, usually through budget cuts – or changes in project specs – or both. In due course, adjustment over adjustment, the final output delivered will be changed, not resembling the one promised at the start.

> End-users could not be more careless about the construction process *per se* for what they really care about is the final output to meet their expectations. Such challenges cannot be resolved until end-users become an integral part of the construction value chain. And so, for the final output to appeal to end-users, they have to become part of the entire process, co-designers of projects and co-creators of value (Mrs. J.N.).

5.1.1.17. *Scheduled timelines left in tatters*

Besides budgeting, time is a metric that is worth considering in large construction projects. The construction process is a broken chain where construction works follow a cyclical curve: we work, we stop, we resume work and then, for some reason, we stop, etc. Often, construction calendars are underestimated because, for instance, they do not account for changes in state-imposed construction standards and codes, which may occur during the lifecycle of a project.

> We often spend lots of time revising the specificities of our projects to ensure they comply with newly-imposed construction standards. Construction calendars do not account for such niceties, which is problematic for construction companies. Theoretically, yes, we could stick to the timelines set, yet, in practice, some unforeseen events may arise over time and sway the timelines of our projects (Mrs. S.P.).

5.1.1.18. *Additional costs incurred to integrate new technology and upgrade assets*

If real estate developers fail to customize their offerings so they meet end-users' real aspirations, they would sooner or later reach a critical point where there would be no more demand for their propositions. Should real estate developers keep their production systems rolling as usual, no successful pitches would be expected to be made – and stocks of unsold units would most likely get larger over time. Besides, end-users who think they have bought properties in smart cities, but were in reality deceived, would soon enough find themselves paying extra money to upgrade their assets and integrate new technology into their homes.

5.1.2. *Social*

We herein list and explain the social problems in construction. First, we tackle off-balance or uncontrollable problems and then move on to controllable ones.

5.1.2.1. *Governments' reticence to share data (O-B)*

Big data is compulsory for a country like France to be able to go farther than building eco-districts to connecting neighborhoods and cities. Having such sizable data at hand, we would be then able to do a number of things with it (e.g. create apps, analyze data, create simulators and others). Actually, immense use-value could be extracted from such data.

> In France, we are still lacking big data at the level of the construction industry. By, for instance, gathering energy consumption information

> written on electricity bills and entering them into a unified database and then proceeding by technically analyzing it, we might find practical ways to enhance the energy consumption and efficiency of our buildings (Mrs. A.T.).

Indeed, in the realm of smart cities, data is of great importance, and, we say, sizeable data could not be secured unless it was exchanged among actors, mainly public authorities and city residents. Through back-and-forth exchanges, data would eventually, via a snowball effect, grow in size. So would the value that may be pulled out from it.

> What is inopportune today is the reticence of some governments around the world to share data with the public which holds them back from building their own smart cities. Such problems are less frequent in democracies in contrast to other political (non-democratic) systems (Mrs. A.T.).

5.1.2.2. *Misconception of smart city notion*

There is no common definition for smart cities, and a lot of misconception is still spinning around the expression today[16]. At present, the subject is very fashionable and professionals all around the world seem to have adopted it, each with their own meaning and adapted it to their own work field. Nevertheless, the topic remains very indefinite as we do not have a shared vision of it.

> Whenever one talks about smart cities, he directly thinks of collective well-being – as the two are strongly connected (Mr. D.T.).

In urban terms, we avow, the smart city concept means nothing, especially in France.

> There is nothing smart about the constructions and structures we see walking down the streets of our cities today, though they are often labeled as such (Mr. D.T.).

Though a smart city is nothing but a structuring axis of development, the main issue with it lies in its longevity. For construction actors, it is often hard to keep a promise that we make today 10 or 15 years later.

---

16 Smarter City myths and misconceptions. Available at: https://theurbantechnologist.com/2013/07/14/smarter-city-myths-and-misconceptions/ [Accessed June 21, 2021].

> Mobility and livability might be the main components of smart cities today, but this might not hold true tomorrow (Mr. F.C.).

Therefore, we proclaim, smart city is an expression, a unifying term, that we often use to embody places where people can assemble and live together. At the end of day, the city is a human construction rather than an architectural or urban one. Thus far, we cannot conclude that smart cities have failed in France – as some promising signs of urban developments are appearing on the horizon. However, we assert that what is missing today is clearly the human factor. Thus, by integrating residents into construction processes, future smart city initiatives would have better chances of being successful.

Nowadays, construction companies are focused on the building: they design, build, promote and sell. What is being sold is the performance and the quality of the building. Yet, whenever we tackle the smart city concept, things change a bit. If we think on a wider scale, say, for instance, on the scale of a district, or that of a neighborhood or a city, then what would be built and sold goes beyond mere constructions to reach social connectedness and well-being. As previously noted, there is a growing tendency among construction companies today to label as smart any project that includes some form of technology.

> Actually, smartness has become a mere marketing tool that holds no meaning that companies habitually use to sell their merchandises faster and for a higher price (Mr. B.G.).

5.1.2.3. *Outdated customer value proposition*

People's interests are often very divergent – and their preferences might change depending on several factors (e.g. professional occupation, social standing, education, etc.), thus the need for construction companies to diversify their offerings so they can fulfill various market demands.

> To draw all end-users together and put them into one category as if their needs were similar is a mistake that construction companies often commit (Mr. E.N.).

For that reason, it is vital for construction companies today to ask themselves: who their customers are: merchants, residents, tourists, small businesses or large businesses? This is crucial as construction is not solely focused on the residential, but extends to the commercial, too. Anyhow, we ask: would a city be deemed smart or attractive if it does not attract residents, tourists, businesses, commuters and others?

> For a city to flourish, it has to become an attractive destination for all customer segments, not just one (Mr. E.N.).

The 4 Quai de la Seine in Paris would have had zero value, if it did not have the ability to attract tourists, capture investments, and strengthen the economic engine.

> Smart cities are conglomerations that entice everyone; a smart city should be human-centric, built for and with the people (Mr. E.N.).

Consequently, people should partake in the construction process of smart cities, not in the sense that they would participate in the construction works, but in the co-creation of value at city level. Moreover, construction companies have to come up with new offerings as people are fed up with spending hours on the roads to get home or to work. Therefore, we trust urban planners to have a crucial role to play in this respect, mainly by working on adapting urban spaces to people's exact needs.

> It is for and foremost a question of making better use of spaces (Mr. A.).

Whether in France or elsewhere in the world, the influx of people towards cities has been pointedly increasing over time. In China for instance, Shanghai has been a migratory area hosting over a million new inhabitants every week.

> People get richer by being in cities. Though smart often denotes what is chic and new, it should not be distanced from the social – as the social is an integral part of the smart (Mr. D.B.).

Often, construction companies tend to mistakenly misjudge people's real needs; they offer the most complex and intricate, whereas people are often on the lookout for the simplest and the most basic.

> Smartness is not about complexity apart from about simplicity; it is about what people want, their well-being, and finding ways to improve the quality of their lives (Mr. D.B.).

5.1.2.4. *Banishment of any collaborative urban governance scheme*

Smart cities, we assert, cannot be built but through collective thinking and intelligence[17]. At the end of the day, smart cities are the result of collective rather than individual effort. Indeed, it is not about Bouygues or any other big operator – it

---

17 Effects of collaboration in projects on construction project performance. Available at: https://www.diva-portal.org/smash/get/diva2:1008892/FULLTEXT01.pdf [Accessed June 21, 2021].

is not even about the GC or the architect – but it is about an amalgam of construction actors who should synchronize their works and collaborate with one another to create the city of tomorrow. A smart city is thus a platform where value creation is a function of value networking.

> A new GC BM for the building of smart cities is definitely an interesting idea, especially for large companies as they are the ones that could espouse it and market it. It is an investment that SMEs cannot afford (Mr. C.R.).

> Almost 95% of construction companies operating in France are SMEs with the bulk of them lacking the required means and knowledge for the construction of smart cities. They are not literally unknowledgeable, it is just that the world is evolving at a very fast pace – and if these companies were not given some sort of grace period to be able to catch up, that is, adapt and enhance their skillsets, all smart initiatives they undertake would most likely be destined to fail (Mrs. A.T.).

> In France, collectivism is almost non-existent. This is forthright as people are often inclined to work individually – and whenever they work in teams or as part of a group, say, in an endeavor to achieve a common something, they often come to disagreements, they face problems, they fight, they fail (Mr. B.G.).

5.1.2.5. *End-users not involved in the early stages of construction*

Almost 10 years ago now, Hermitage Construction, a Russia-based company, was commissioned to plan and build two towers each of an altitude of 320 meters in La Defense, next to a housing complex that dates back to the 1970s. Once the project kicked off, hundreds of inhabitants living in the project's environs left their homes. The project has been in progress for eight years – and now it seems on the verge of being cancelled. Though elected officials as well as other public authorities have worked together with local communities on the project, it looks clear that the discussions held were not genuine. This is to say that construction companies cannot presume they know what people really want because they simply do not. Today, besides consulting with potential clients, construction companies must work on diversifying their offerings so that what is being offered, a variety of propositions rather than one, acting as a net, ends up enticing the largest number of customers possible.

> By contemplating the newly built structures in the city of Paris, one could clearly see that the output being delivered is far from being smart. In smart projects, it is not a building or an asset that we are

> selling but a lifestyle, a feeling of well-being, and a quality of life. The author's GC BM proposal is very interesting especially in that construction companies are now invited to revise their somewhat outdated value propositions (Mr. B.G.).

Today, real estate developers seem to be unaware or disconnected of the real world, of people's real needs. In the smart city arena, France is lagging behind relative to its neighbors, mainly Spain and England – and some serious efforts must be deployed for the country to be able to restore its position in this sphere.

> When the project is not co-conceived and co-designed with local communities (potential end-users), the odds that conflicts would arise once the project has taken off are expected to be high. It is an inherent risk to large-sized construction projects, a risk that is often underestimated by both public authorities and industry actors (Mr. F.C.).

5.1.2.6. *End-users not being co-creators of value*

Smart cities are mainly a question of finding ways to keep information circulating and alive. This is crucial, we proclaim, as information tends to become diluted over time. The longevity of smart developments poses some serious problems for construction actors, the reason why finding resourceful communication canals or a system where data can flow freely and openly and be archived seems highly necessary. Despite the fact that we may communicate lots and lots of information to people about the project being studied, the odds they will forget most of this information are high. And so, to keep end-users involved throughout the entire construction process, irrespective of its length, is a crucial thing for the success of smart developments, a means for boosting social acceptability.

> By allowing end-users to act as construction partners, the risk of end-users complaining about the project upon delivery could be significantly reduced (Mrs. S.P.).

5.1.2.7. *Criticism of final product by end-users*

Often, end-users are less concerned about the technicalities of construction projects and more concerned about the quality of the final product they would be getting[18]. As previously noted, France-based construction companies are facing many challenges in operating, executing and managing large construction projects.

---

18 Construction Firms: Seven Essential Tips for Client Communication. Available at: https://www.workflowmax.com/blog/construction-firms-7-essential-tips-for-client-communication [Accessed June 21, 2021].

Among those, we cite a few: budget and time overruns, mismatches between promised and delivered products, absence of after-sale services, tech-intensive constructions, poor urban design and others. All of these challenges combined will eventually be indirectly passed on to the end-user whose journey would be negatively impacted. He would be dissatisfied with the product purchased and have a lot of complaints about and concerns regarding it. In some cases, end-users take the necessary legal action against developers from whom they purchased the property.

5.1.2.8. *Broad concerns a propos the liveability of assets*[19]

Once the final product has been delivered, should end-users' assets require maintenance works, they would most likely find themselves alone, as no construction actor would plead guilty and accept to assume the costs that could ensue from such low-quality works. At the end of the day, end-users would be totally discontented. And their indisposition to recommend the services they received to others would ultimately (negatively) influence the business activities of the construction companies concerned. For those companies, the impact of a low NPS (Net Promoter Score) would not be felt until after a while – in the mid to long run.

5.1.3. *Environment*

We herein list and explain the environmental problems in construction. First, we tackle off-balance or uncontrollable problems – and then we deal with controllable ones.

5.1.3.1. *Urbanization (O-B)*

The need to adapt knowledge and skills was only felt recently, after the global surge of the smart city concept. The rise of smart cities, we proclaim, is the direct result of urbanization, which has lately become a trending phenomenon.

> The existing know-how among construction actors in the French construction industry is somewhat old and requires substantial improvement. Urbanization is a function of technology and society, and the processes of technological and social changes are functions of urbanization (Mrs. A.T.).

Though most people cannot see anything but the negative effects of urbanization (e.g. overloaded infrastructure, destruction of habitats, overcrowding, pressure on

---

19 Spotlight on livable cities, Part III: Responding to the challenge of livability. Available at: https://www.urbanet.info/livable-cities-challenge-of-livability/ [Accessed June 21, 2021].

ecosystem, etc.), it is worth noting that some noteworthy benefits could also ensue from it, with the most obvious ones being: economic growth and development, and the increasing need to innovate and improve existing production systems.

5.1.3.2. *Constantly changing construction codes and standards (O-B)*

The constant changes to state-imposed construction codes and standards are going much faster than what construction actors could possibly accept. In France, there is a very high turnover in terms of the standards enforced. Indeed, I say, construction standards are often experimental with a life span that does not exceed two years.

> A project is launched in the morning according to specific standards, yet, in the evening, new ones emerge, meaning that we have to redo everything from start. The rhythm based on which construction standards are changing is very fast (Mrs. A.T.).

This is quite challenging for construction companies. Though these recurrently changing standards might be imposed to protect people's interests and serve their needs (which they might!), I trust that people care less about the standards and more about the quality of the final output and the amenities accompanying it. People seek comfort above all, and amenities play an important role in offering quality living.

> The reference frame of state-imposed construction standards is often detached from people's day-to-day needs (Mrs. A.T.).

Energy performance is among the key topics that springs to our mind when thinking of environment-related construction standards. To my knowledge, energy performance, say at building level, could be achieved in so many ways and using different technologies and systems. In fact, I could openly assert that solar water heating systems is the main technology (to my knowledge at least!) being used by construction actors today to enhance the energy efficiency of their buildings. The said system is relatively cheap and its use-value can be easily grasped – and so, people are often willing to pay for it. Yet, when it comes to other more advanced technologies and systems, people seem to be out of tune.

> There is a huge gap between construction standards and people's preferences, with the former changing at a much higher speed than the latter (Mrs. A.T.).

5.1.3.3. *Lack of smart city initiatives and urbanization efforts*

Moving on, I, as an expert in the French construction industry, can openly avow that what people are actually looking for does not really come hand in hand with what is being offered. A new concept that I think is worth putting emphasis on is that of urban villages. In China, urban villages, though overcrowded, are among the country's liveliest areas, notable for affording economic opportunity to newcomers. They constitute a unique phenomenon that have formed part of China's urbanization efforts. Today, in France, some similar initiatives must be implemented for the population to be able to bring transformative change to French cities. Though the number of national programs in favor of smart city initiatives being run by different government bodies has been on the rise in recent years[20], a lot of work remains to be done for the country to become a leader in the smart city market.

> In France, initiatives such as building eco-districts are good, yet insufficient – as those account for merely 5% of the country's total urban structures built. Today, we are still figuring out how to build the new city over the old one (Mr. P.B.).

5.1.3.4. *CSR-related topics occasionally discussed for mere marketing purposes*

Corporate responsibility strategies have been gaining increased momentum and attention from experts operating in the construction industry in recent years as questions of sustainability and green agendas have come to permeate business the world over. Construction companies may not like to openly disclose it, but CSR (Corporate Social Responsibility) is indeed part of their marketing strategies to build a positive brand image among customers.

Indeed, I believe that customers are more willing to deal with construction companies that have a good reputation in the marketplace. Concisely, CSR provides construction companies with the opportunity to identify and address the harms they are contributing through commercial activities to society. By identifying the right social cause and addressing it, construction companies will be able to identify themselves as responsible and trustworthy. However, the problem today lies in the fact that most French construction companies still do not understand the most effective ways to design and implement sustainability programs, which explains their incapacity to fully capitalize on its potential for the creation of business value. Time and again, construction companies use CSR as a mere marketing tool, an ornate paper they use to wrap their outputs so they can be commercialized in the marketplace with no trouble.

To be precise, to date, they have mostly focused their attention on direct routes to extracting business value from CSR. They are pursuing easy-win strategies that have direct commercial benefits, such as computing and reducing their corporate carbon

20 The French Tech project, the Eco-quartier Label project, and the Investment for the Future project – to cite but a few.

footprints. Though such strategies are deemed useful, they often fall far short of the mark. In fact, for construction companies to be able to fully benefit from CSR, they must wake up to the fact that they need to take a more indirect route to creating value with it. Practically, this involves moving away from top-down managerial logics in the direction of richer bottom-up ones where value could be co-created in partnership with other stakeholders. Indeed, we ask: should construction companies sacrifice the environment or their profit margins? Honestly, there is no immediate answer to this question. To date, we can say, construction companies are still focused on achieving short-term financial goals, that is, maximizing revenues and minimizing costs. In their belief, if effectively applied, CSR would end up being high-priced and could possibly cause their financial proceedings to plummet. Today, construction companies are invited to find a certain balance between the economic and the environmental aspects of their projects (a vital matter in the context of smart cities!).

> The environment is always regarded as an optional aspect that will be honored if and only if the project's budget permits it – otherwise, whenever cost cuts are necessary, it is frequent that the environment will be sacrificed so that the economic aspect can be preserved (Mr. F.C.).

5.1.3.5. *Non-sustainable construction*

In France, buildings are responsible for the bulk of energy consumption and GHG emissions[21]. Indeed, residential buildings, constituting about 75% of the total floor area, mostly occupied by single-dwelling buildings, dominate France's metropolitan building stock[22]. While population and floor area are expected to grow in the near future, final energy demand for heating, hot water and ventilation (among others) is estimated to drop by a few points. This is mainly the result of the newly imposed building code requirements. On the supply side, though key stakeholders trust the use of RES (Renewable Energy Systems) and in-site productions could play a vital role in boosting the energy performance of buildings, we openly assert that the use of RES on a national scale is currently impossible, especially with the present energy prices, making RES an investment that is hard to pay off.

> Whenever building smart, the structures themselves should be smart and eco-friendly, built using sustainable materials. However, such materials are often expensive which could pointedly impact construction companies' profit margins. Often, companies refrain from building sustainable constructions for purely economic reasons;

21 Building Market Brief (France). Available at: http://cuesanalytics.eu/wp-content/uploads/2018/10/181023-CK-BMB-BMB_FRANCE-DEF-CIE-Edition.pdf. See also: https://www.archires.archi.fr/en/catalogue/90.

22 Ibid.

> and constructions will remain non-sustainable in France until they become economically sustainable first (Mrs. S.P.).

5.1.3.6. *Large amount of logistics, pieces and other resources tied up with the process*

Usually, construction projects go through different stages from initiation to completion. Concerning lengthy smart developments, their execution requires a great amount of logistics, which makes them creators of negative externalities.

> The orthodox procedures based on which construction companies are operating these days may be improved by bringing some humble yet valuable variations to the existing value chain (Mrs. S.P.).

Today, the focus of France-based construction companies is on the building of eco-districts – which we believe to be sustainable rather than smart constructions. Though this might still look good, these eco-districts are not always built following eco-friendly procedures. Complying with state-imposed environment-related standards and codes is certain to reduce the financial returns that we could expect to get from a project. Thus, construction companies often make concessions at the expense of the environment in an attempt to prêserve a decent profit margin at the end of the construction process.

> Often, the economic prevails over the social and the environmental (Mrs. S.P.).

5.1.4. *Technical*

We herein list and explain the technical problems in construction. First, we tackle off-balance or uncontrollable problems, and then deal with controllable ones.

5.1.4.1. *Passive bidding processes (O-B)*

Large construction projects are often the chattels of public authorities who carry out passive bidding processes for the assignment of such projects. It is the lowest bidder that is frequently selected for the job. Generally, a truncated price is a sufficient criterion for a company to win the project, regardless of the technicalities of the project and whether the selected company has a pertinent history in the handling of such projects (or not).

> In France, construction standards are often set in favor of specific construction actors who seem to be always protected by the law. In addition, when it comes to public procurement procedures (tenders), it

> is evident that big operators such as Bouygues, having armadas of lawyers and, as a result, developed bargaining abilities, are better positioned than their smaller counterparts to win such bids (Mr. B.G.).
>
> Within this framework, public tenders issued, often miswritten and lacking key criteria that bidders should account for in their proposals (but do not!) ensue, once the project is commissioned, in ample operational issues (Mrs. S.P.).

Overruns are not serious issues *per se*, as the specs of the project can always be revised so they fit the project's budget envelope. Indeed, serious issues are only felt when erroneous estimates are significantly large so that no adjustments could be made to correct them. Henceforth, any potential tune-ups in such a case would be made at the detriment of other project specs.

> Construction companies draft their technical and financial proposals based on the specificities of the project, exactly as mentioned in the terms of reference. They have a whole series of solutions to opt for whenever operational difficulties are faced, yet we can't expect miracles when the project owner himself is unknowledgeable of the most basic components of smart developments (Mrs. S.P.).

5.1.4.2. *Legal and institutional barriers (O-B)*

The absence of a national policy and a dedicated smart city program for the promotion of multi-dimensional urban development is the main reason that has been pushing French cities, alone or in clusters (OASC France)[23], in partnership with private technology providers (EDF, Veolia, SNCF, etc.)[24], to become proactive and undertake personal initiatives for the smartization of their territories.

> The recently imposed Élan law failed to solve the numerous urban governance issues encountered today in the French construction industry. It is the French authorities' conservative standing vis-à-vis innovative initiatives in construction that is holding the entire industry back from innovation (Mr. D.B.).

23 The OASC France (Open & Agile Smart Cities France) initiative is a city-driven non-profit organization with the aim to create a Smart City market. OASC France was initiated in 2015 by four French cities in the north of France (Communauté d'Agglomeration de Saint-Quentin, Valenciennes Métropole, Amiens Métropole, Ville d'Arras).

24 Smart eco-cities in France: Trends and city profiles 2017. Available at: http://www.smart-eco-cities.org/wp-content/uploads/2018/07/Smart-Eco-Cities-France-2017.pdf [Accessed July 6, 2021]. More details on construction laws available at: https://construction.laws.com/general-contractor.

Another issue that is worth noting at this level relates to the fact that smart city projects are often badly conceived, handled by inept stakeholders who fail to manage them properly.

> City mayors are part of the problem. Though they are unskilled in construction, with the texts of law, absurdly, not requiring them to be, they often decide on the designs and shapes of the structures to be built. What we need today is an innovative urban governance model for the building of coherent and resilient smart cities (Mr. E.N.).

5.1.4.3. *Smart city projects as disconnected micro projects*

In France, policy makers should trust in competition as a means of enhancing quality and sinking prices. Ten years ago, the works of real estate developers revolved around public developments; they had a wider vision of the city (as an urban agglomeration); they used to think of the city as a platform that should be built harmoniously; nonetheless, today, they are inclined to operate on a project-by-project basis, with projects built disjointed from one another.

> A smart city is a platform where different smart projects are built and put in connection with one another. It is the connectivity of all projects within a city that makes a city smart (Mr. B.G.).

5.1.4.4. *Too much weight put on technology in the building of smart cities*

The banal conception of smart cities is among the key problems that the construction industry is facing today. This is accurate, as big operators, repeatedly, tend to connect smart cities with the simple act of installing small electric meters (for instance) at the building level. As we see it, a smart city is a destination rather than a means.

> It is the incompetent processes being used nowadays by construction companies rather than state-imposed legislations that are putting brakes to the development process of smart cities in France (Mr. E.N.).

Thus far, industry actors are still misconstruing smart constructions for technology-intensive ones[25]. In the context of human-centric smart cities, this is exactly where the problem lies.

---

25 When to Launch Technology and How to Stick the Landing. Available at: https://www.constructormagazine.com/virtual-design-and-construction/ [Accessed July 6, 2021]. Other useful insights on the matter available at: https://constructech.com/news-brief/.

> Technology is a means rather than an end; it is a tool that should be resourcefully used to connect all constructs of a smart city while ensuring resource optimization (Mr. P.B.).

Ideally, smart cities are built following a pragmatic managerial approach where available tools, resources and skills are intelligently exploited to address the challenges faced in construction today. We close by affirming that smart cities cannot be constructed but through trial and error, crafting apposite urban governance and resource management schemes, ensuring interoperability between stakeholders and enabling technological innovation and systemization.

5.1.4.5. *Overlooked organizational innovation*

The global success of the United States is mainly driven by entrepreneurial initiatives, the creation of start-ups and technology which is often put at the service of both businesses and people. Singapore, we say, is a good example of a country that has succeeded in building its own economic model by copying other developed countries' models – while adapting them so they fit its own form. Indeed, by espousing innovative production systems, Singapore was able to improve the urban design of its cities. As previously discussed, there is no perfect business model for the building of smart cities, as the meaning of smart cities can conceivably change across countries and cities, and so would the BM crafted for their building. In France, we say, the construction industry seems to have failed where the aeronautics and automotive industries thrived.

> We have managed to fly planes with pieces manufactured in the four corners of the world without reporting any problems throughout the process – however, now, in the construction industry, we seem to be encountering copious challenges building a simple structure around the street corner (Mr. C.R.).

While Paris is still lagging behind today, other nearby cities, Barcelona for instance, are really ahead of the curve. Spain, being innovative in construction, has succeeded in turning Barcelona into a smart metropolis following a people-centered governance model within which technology (BIM) has played a vital role[26].

> There is no magic wand for the building of smart cities. A smart city could not be built but through trial-and-error, a recurrent process, where we build, we adjust, we rebuild, we upgrade, so on and so forth, until we reach a situation where the final output has been sufficiently

26 Information on global smart city constructions – and management practices are available at: https://www.constructionglobal.com/infrastructure/smart-cities-need-smarter-project-management.

> upgraded and improved to then be labeled smart. Les-Bricolos-du-Dimanche, is a garage-based small trade whose owner never imagined that one day he would become a pioneer in a craft that was initially inexistent (Mr. E.N.).

Today, a lot of flaws and brakes should be tackled at the level of the construction industry. We should add that a ready-to-use solution is simply unavailable. Moreover, despite so much talk about the Parisian smart city, no serious measures have been implemented in this regard – which explains why the city remains non-existent to date. As previously noted, what is missing today is neither the technology nor the know-how, but a fitting governance model that would allow for the resourceful building and management of smart cities. Hence, we say, the problem is chiefly organizational, neither technical nor technological.

> Smart cities are coherent ecosystems that should be built and managed by maven entities. Construction is never always about the technical and the technological, but about the organizational too. Innovation in construction should not always be radical as small and incremental changes brought to the value chain could time and again be sufficient to resolve arising problems and restore the chain's ability to create value for all (Mr. E.N.).

5.1.4.6. *Smart constructions as pure technology demonstrators*

The smart city concept has nothing to do with creating green buildings, but with crafting something that could fit into a global ecosystem[27]. In France, public developers presume they know how to build smart cities. They build structures and invite people to come settle in them. This does not seem like a bad idea; however, the main challenge lies in their capacity to convince people to come and invest in those projects. It is a question of social acceptability, creating a rapport between people and the structures built. If such a rapport is impossible to create, then the structures or cities built would end up being empty.

> The city *per se* is nothing; it is just a bunch of walls and streets. Smart cities must be friendly, digitized, and agricultural; they should also appeal to people – be human-centric – account for urban mobility, etc. – otherwise, they would not be smart (Mr. C.R.).

Though smart cities are often built using technology (referring to digital infrastructures)[28], technology *per se* should only be regarded as a mere tool, a means used for a specific purpose, that is, the building of smart cities.

---

27 More details on the matter available at: https://constructech.com/news-brief/.

28 Ibid.

> To get to the Rue-de-l'Avenir Association, a pedestrian has to cross four lanes, two for buses and two for cars. Similar urban snags could be bumped into in Tour-de-Montparnasse as well as on the T3a-T3b junction at the Porte-de-Vincennes. Such urban planning is neither intelligible nor intuitive. We have been told that our cities are smart, yet in reality, there is nothing smart about them. Sometimes, our cities are not even humane (Mr. E.N.).

> Mismatches between conception and execution plans – and the absence of a strategy that could oversee the execution of large projects are possibly the two main causes of such urban design flaws (Mr. D.B.).

> How did Romans, roughly five thousand years ago, succeed to design aqueducts and other complex structures without the use of technology? What is astonishing is that the Romans made the trajectories of tunnels through alignment with the reflection of light, and they achieved their works flawlessly, from start to finish, without any technological assistance. Also, to make the first Airbuses, we had one part manufactured in Germany, a second one in France, and a third one in the UK – and finally, all parts were put together in the US using made-in-China machineries. Again, we used to fly planes with materials manufactured by people who neither spoke the same language nor had the same cultural backgrounds, and all of this was done faultlessly – whereas, in our time, with all the technology we have got, we face lots of problems building a few houses around street corners. The problems faced today are thus governance-, not technology-related (Mr. E.N.).

5.1.4.7. *Risk of technological obsolescence upon delivery of project*

The lifecycle of a smart construction is quite a bit longer than that of a technology. Therefore, whenever a smart construction project is initiated, and by the time construction works are complete and the final product is ready for delivery, the odds the technology implemented is obsolete are high.

> All people who thought they had bought smart homes five years ago are finding themselves today paying significant amounts of money to upgrade their properties. We build over a 24-month period but, over that period, we are at risk of overlooking two generations of technology products which makes our construction archaic at delivery. Besides, once a technology has been implemented at the very early stages of the construction process, it would not be feasible, cost wise, to simply confiscate it and implement a new one (Mr. V.B.).

CONSTRUCTION STAGE	*PROBLEMS*			
	ECONOMIC	***ENVIRONMENTAL***	***SOCIAL***	***TECHNICAL***
INITIATION & CONCEPTION	*– Failing BM leading to abridged growth prospects at industry level – Lack of understanding of smart construction process – Top-down managerial approach – Big companies' resistance to change – Financing (O-B)*	*– Urbanization (O-B)*	*– Misconception of smart city notion – Banishment of any collaborative urban governance scheme – Governments' reticence to share data (O-B)*	*– Passive bidding processes (O-B) – Legal and institutional barriers (O-B) – Smart city projects as disconnected micro projects*
	– Constricted approach to value creation	*– Lack of smart city initiatives and urban efforts – Constantly changing construction codes and standards (O-B)*	*– End-users not involved in the early stages of construction*	
PLANNING	*– Project process confused for business model*		*– End-users not being co-creators of value – Outdated CVP*	
	– Faulty estimates leading to recurrent budget overruns			*– Too much weight put on technology in the building of smart cities – Overlooked organizational innovation*
	– High number of stakeholders with overlapping roles and responsibilities	*– CSR-related topics occasionally conversed for mere marketing purposes*		

LAUNCH & EXECUTION	*– Decentralized decision-making* *– Discoordination between construction actors*	*– Non-sustainable construction*		
	– Innovation mismanagement			*– Smart constructions as pure technology demonstrators*
	– Fragmented projects	*– Large amount of logistics, pieces and other resources tied in with the process*		
MONITORING & CONTROL	*– Loss of control over quality of works*			
CLOSURE	*– Mismatch between promised and delivered final product* *– Scheduled timelines left in tatters* *– Absence of an entity that could orchestrate time horizons and handle maintenance works*		*– Criticism of final product by end-users* *– Broad concerns a propos the livability of assets*	*– Risk of technological obsolescence upon delivery of project*
RESTORE, RENOVATE, REMODEL	*– Additional costs incurred to integrate new technology and upgrade asset*			

Table 5.1. *Key problems in construction (France) (source: created by the author)*

Table 5.1 summarizes all problems discussed throughout the current section. They are tabulated based on *stratum one and four* of our GC BM, by construction stage and distributed across the four layers of the TLBMC.

The next section embarks upon the plausible solutions for the problems identified, except for those designated as off-balance.

5.2. Solutions

Up until now, we have made it clear that the French construction industry is currently facing countless difficulties handling and managing smart developments. In addition, we have put an emphasis on the fact that the construction industry has been for a few years now operating based on a failing BM, which, apparently, has been slowly but surely seen its performance and productivity plummet. We have also showed that, not only in France, but also around the globe, there is a lack of a straightforward and precise definition of smart cities. Indeed, as we have said, the smart city concept is highly subjective and may possibly hold distinctive meanings among its layers. In fact, smart could mean anything, as anything could be smart: smart grids, smart constructions and smart buildings – just to cite a few. In France, there is still confusion today among stakeholders about what smart cities entail and what they should consist of.

For my part, “a smart city is a city that is conceived, developed and run by the government and the people, both being smart, for the sake of promoting citizens' well-being and ensuring a better quality of life for all”.

Though this definition might sound good, it does not really help, practically at least, in deciphering the key constructs of a smart city, how the latter should be built, what it should resemble or even what it should encompass. The smart city concept is multifaceted and extends to reach anything and everything, the economic as much as the social, the environmental, the technical and even the political. This last sentence may sound a bit philosophical, yet, based on the extensive readings I have done in this regard, this is what smart cities really are. Here and now, what we know for sure is that there is currently no one perfect, all-inclusive smart city model because, as reported earlier, smart cities are a function of several, country-specific, builds. In other terms, smartness is likely to change meaning based on one's geographic location – that is, based on where a person is. Smartness may denote the simplest of things, as it may also refer to the most complex. With that said, a smart city is never associated with only one construct, but several interrelated ones, which, once pooled together, will find a coherent system that is smart itself. Moving on, we have also affirmed that a lot of money should be invested for the conversion of a big city into a smart one. It is a long-standing, capital-intensive investment.

> Our reasoning comes hand in hand with the declarations of Mr. E.N. who specified that "a smart city is a combination of two elements: financing, and a clear vision of what a smart city should consist of and what its purpose should be. A city would not be deemed smart unless the services it offers come in line with people's exact needs and aspirations" (Mr. V.B.).

Often, construction companies misapprehend the real needs of people. They presume to know what people really want – but, in reality, they do not. This is true as, sometimes, what construction companies believe to be bad ends up working just fine, while what they trust to be fine turns out to be bad. Building a smart city, we acknowledge, is principally a question of abridging the growing market gap between supply and demand. In this book, we focus on a specific construct of smart cities: smart constructions, being structures and buildings linked together through technology and data. In smart cities overall, technology is a means rather than end. It is a tool used to facilitate communication and data exchange among the various elements of a city.

> Smartness is never solely bounded to technology; it extends to cover the social, the environmental, the livable, and the humane within a city. Today, French-based companies are operating based on a pure money-making logic, an inhumane business model (Mr. D.T.).

At present, we do not actually have to rebuild the entire world, as it already exists; all we have to do is to improve it by bringing some worthwhile change to it.

> Smartness should not be mistaken for the complex and the compound, as it frequently relates to the simplest and the most obvious (e.g. create a parking app that people could use to find alternative parking spots, boost buildings' energy performance, implement urban security measures, enhance urban mobility, and others). The buildings we see nowadays walking down the streets of our historical cities are not smart, but are labeled as such anyhow (Mr. D.T.).

For construction companies, projects labeled smart are thought to sell better and quicker in comparison to the ones that are not. This nevertheless creates a misconception among citizens, leading them to think of smart cities as unsympathetic and unlivable places. We regularly tend to complicate things in a misled endeavor to make them smart.

In the following sections, we refrain from discussing off-balance problems as they are out of our reach and control. Accordingly, we solely focus on discussing project management problems and suggesting plausible solutions for their resolution.

5.2.1. *Economic*

We herein list and explain the solutions suggested for the resolution of the formerly listed economic (and other) problems.

5.2.1.1. *Espouse a centralized decision-making process*

Despite the many innovative initiatives implemented at country level today[29], France is still lagging behind in the field of smart city development. This is mainly because large construction projects are often owned and run by incompetent project owners. Being inept in construction, they often outsource, through passive bidding processes, the execution and administration of such projects to external entities, regularly to the lowest bidders. Regardless of the quality and content of technical proposals, bidders who submit the lowest financial proposals are the ones with the highest chances of securing the job. This is problem-zero – from which ensues a series of other problems that will accrue over time and lead to the failure or cancellation of construction projects.

> Large construction projects, smart ones specifically, are hard to manage; they are tricky and lengthy, and the BM based on which construction companies are operating nowadays does not seem to comply with the requirements of a smart city. Today, there is definitely a need for a maven contractor who could step into the scene and use both his knowledge and know-how to successfully pilot and manage construction works carried out in the context of smart developments (Mrs. A.F.H.).

> For smart developments to succeed, construction works should flow smoothly down- and upstream the value chain, and construction actors should be able to flawlessly synchronize their works throughout the entire process, from start to end. Multi-level interoperability must be ensured at all times. For this to be possible, however, we need a central operator, an integrator, someone with adequate expertise in construction to make the right decisions at the right times (Mrs. A.T.).

29 SBA (Smart Building Alliance) is association with offices in Lyon launched to bring change in the field of construction. It brings scholars and industry experts together to discuss and rethink both smart city and smart building notions.

Within the value chain, each specialization has its own *raison d'être*, however, a central operator is needed in order to bring all stakeholders together, thus ensuring interoperability between them. Put differently, the central operator would tie the entire value chain together, ensure a fair interplay between stakeholders and oversee all tasks carried out from initiation to closure. Besides being a decision-maker, the central operator (which in our case is a GC – assigned by the project owner to replace him – acting as a delegated project owner) would be given control over the entire construction process. As an intermediary between stakeholders positioned down- and upstream of the value chain, he would ensure end-to-end data delivery, from the project owner, at one end – to the end-user, at the other end.

5.2.1.2. *Improve city appeal and absorption capacity*

There are precise (inflexible) procedures that France-based construction companies must abide by at all times for the achievement of their projects. Thus, to simply change how things work and jump from the orthodox to the unorthodox would be without a question challenging. Under a smart city business model, resources should be exploited – and services delivered – in a sustainable way. In France, they are not quite there yet. Also, with the rise of the smart city concept in recent years, several key elements that were previously overlooked have come into play (open data, IoT), and new actors have been brought to the scene (ICT companies). The world is evolving and change is part of it. If we fail to adapt, we will eventually lose in business.

> Sooner or later, public governance schemes will change, hence setting the ground for further innovations like the one suggested in this book. Unfortunately, there is no magic wand that we can use to overcome the problems faced in construction today: to fix things, we need to innovate. The city of tomorrow cannot be built but through trial-and-error, an ongoing process where mistakes are made and necessary adjustments follow (Mr. D.B.).

Today, all cities around the globe are aiming to become smart. Smartness is a means that enables cities to boost their allure and gain economic power by attracting capital to their regions. Today, the main challenge consists of finding the perfect recipe for the creation of smart cities, as smart cities may include too little or too much. Generally, smart city projects should be catered based on the exact requirements and hopes of end-users. They should also account for the specificities of the cities or territories on which they are sited. If, for instance, we had to build the same project in Lyon and in Montpellier, we could not expect a comparable success rate in both regions if no adaptations were brought to it *a priori* to reflect the uniqueness of each territory. People in Montpellier are likely to have different tastes from those in Lyon; also, Lyon, as a city, has distinctive traits that differentiate it from Montpellier. Once a construction project is on the go, it should account for all of these minutiae – otherwise it would not be smart.

> Friendly, people-centered, technology-based, data-driven, livable – are key adjectives that one could use to define smart cities. The desirability of a city ensues from the shape and look of the lodging first, then comes the nicety of the shop next door, the view, the amenities and services offered, and finally the greenery available in the vicinities. This is what creates real-estate value and gives a city its real appeal (Mr. B.G.).

5.2.1.3. *Ensure final product has a use-value and that utility may be extracted from it*

No construction actor has control over the entire value chain, which implies that no one can guarantee the quality or state of the final product since, throughout the construction processes, the relay baton is regularly passed on from one stakeholder to another. Based on where a construction actor is located along the chain, his control over the quality or state of the final product is expected to diminish or grow the more we move down- or upstream.

> The prevalent Design-Build-Operate model must be improved so it becomes something like Design-Integrate-Build-Operate. As construction companies, we are not able to keep our promises in terms of delivering a high-quality output at the end of the process simply because we do not intervene at all stages of the value chain. Hence, the need today for a specialized entity who could control construction works done and guarantee that value created throughout the process would not be lost (Mr. B.G.).

At the end of the day, in the context of smart cities, construction companies need to ensure what is being sold is more than simple structures. In collaboration with architects, urban planners and other stakeholders, we trust, the GC would be able to deliver upstream what has been promised downstream. Moreover, construction companies must ensure the final output delivered is useful and utilizable. To be precise, end-users must be able to extract use-value from the spaces built – otherwise they will be dissatisfied. It is mainly a question of space management (use, eco-friendliness, etc.) – as well as of urban design.

> End-users should be all set to use their homes in a convenient and handy manner (Mr. B.G.).

5.2.1.4. *Allow for customizability (portfolio diversification)*

The rapport between architecture and housing has traditionally been an uneasy one. After a period of flagging interest from the 1970s into the 1990s, design interest in housing is again on the rise, predominantly in terms of innovative materials and production systems, green building and an active interest in providing for a broader spectrum of people. The production of housing today ranges from the hyper-customized singular house for a distinct user to completely mass-produced manufactured housing and repetitive building models. Recently, mass customization has become a term used to describe housing production that falls between these extremes. This term suggests a production system that has the stability of quantity and the flexibility of custom design. To be applied effectively in architectural work, mass customization must be understood as a system approach that includes not only design, production and construction, but also communication and economics.

> Today's homebuilders are encountering a production gap between the need for product standardization that helps reduce construction costs and the need for product customizability that satisfies the diverse demands of contemporary consumers (Masa Noguchi)[30].

In other words, construction actors are trying to mass-produce housing in order to create stability and mitigate risk. Simultaneously, more types of consumers are entering the market and more consumers of all types want to have a larger voice in the design of their homes. Bridging this gap, we proclaim, requires a system that is familiar and standardized enough to allow the construction industry to embrace it – while also malleable enough to respond to the diverse needs of end-users at all positions in the economic and cultural spectrum. Architects and producers of mass-customized housing are today attempting to create just such a system in which affordability and quality–diversity–sustainability are not mutually exclusive. This kind of system takes available off-the-shelf components and uses them to create modular building and furnishing elements that could be assembled in a variety of configurations. In fact, this suggests something more egalitarian, available to more people.

> Customized housing is a bottom-up approach that comes from the assembly system itself rather than a top-down approach used in more conventional prefab construction (Kieran Timberlake)[31].

30 Mass-Customization in Housing: Designing Systems Rather than Objects. Available at: https://lib.dr.iastate.edu/cgi/viewcontent.cgi?article=1094&context=arch_conf [Accessed July 11, 2021].

31 Available at: https://kierantimberlake.com/.

Nowadays, the increasing competitive pressure on construction companies requires the revision of existing corporate structures[32]. Accordingly, extending product ranges seems to be one plausible way that construction companies could opt for to secure their market positions in the long term. Another possibility is to strengthen the core business. This ensures the function as price or quality leader. Due to the lack of financing, the realization of new (large) construction projects has become more difficult today. The lending of many banks has become more stringent, and many companies are pushing their construction projects to procrastinate. To compensate for the prevalent unstable situation, a trend towards diversification has been recently apparent and the decline in sales was somewhat offset in this way. Additionally, the possibility to operate in different segments of the construction industry compensates for the risk of cyclical fluctuations. Large, international corporations, in particular, are active in activities along the value chain. One of the advantages is that even large projects could be managed almost entirely from a single source. Generally, there are two possibilities for diversification: vertical (the inclusion of upstream or downstream construction services in the product range) and horizontal (the expansion of the product range to include services or products, which are closely related to the current performance spectrum) – and it is up to each company to choose which of the said possibilities suits it best.

5.2.1.5. *Identify and target underserved market segments*

There are various ways to determine if a company is innovative or not. For some, innovative companies could be new entrants that serve previously unmet needs. For others, they are those that introduce groundbreaking solutions to relevant issues. They could also be businesses who use strategic armaments such as technology, design or sustainability to create an impact. These are all in various lists of the top innovative companies that aim to help underserved niche markets (e.g. Nespresso). As mentioned, most corporate innovations focus on markets with certain similarities. First, there are minimal logistical challenges. Then, there are consistent BMs in the same competitive sphere. Lastly, they return a timely profit greater than or equal to other investments. Meanwhile, underserved markets present greater challenges. Consumers in such markets face often overlooked issues. Resources are insufficient, and outcomes are not always predictable. For some innovative companies in this space, these challenges represent unseized opportunities. Indeed, their measures for success deviate from the usual practices. In France in particular, social housing, for instance, represents an opportunity for construction companies seeking to diversify their offerings and enlarge their customer bases. A book – *The Fortune at the Bottom of the Pyramid* (Prahalad and Hart 2004) – proposed that 4 billion people

32 Market positioning through diversification. Available at: https://buildingradar.com/construction-blog/diversification-construction-industry/ [Accessed July 11, 2021].

earning less than US$1,500 a year represented a US$5 trillion global market. While individual incomes in underserved markets are low, the aggregate buying power of communities remains huge[33].

5.2.1.6. *Adjust offerings to exact needs and requirements of end-users*

Construction companies are invited today to offer tailor-made solutions to their customers. This is genuine as not all people have the same tastes and preferences. While some might prefer the old and the traditional, others would be more fascinated by the innovative and the new. This solution is directly linked to the ones mentioned above, primarily, customizability and portfolio diversification.

5.2.1.7. *Ensure involvement of end-users in all stages of the construction process*

Nowadays, at the very early stages of construction processes, construction companies tend to make concessions with customers to ensure their expectations and needs are imprinted in the projects. Though this might not be a common attitude among all construction companies, things seem to be gradually shifting this way.

> Concessions are not always serious and real, sometimes arranged because they are simply part of the process. Passive concessions made at the beginning of the process would eventually ensue in serious clashes with customers at the end, upon delivery of the final product (Mr. P.B.).

That is, value propositions made irrespectively of people's real needs and aspirations would be destined to fail. Even though such propositions were successfully commercialized once, deceived end-users would make sure, through negative word-of-mouth, that such an eventuality would not occur again.

> Urban design has a substantial impact on people's daily lives and so construction projects should not in any case be conceived and built without consulting with potential customers first. The final output delivered should be tailor-made to their exact needs and construction procedures should be adapted so that such offerings could be made (Mr. F.C.).

Between yesterday and today, client integration has made a complete difference. End-users have become (or are on the way to being!) co-conceivers of construction

33 Top innovative companies that target underserved markets. Available at: https://www.boldbusiness.com/society/underserved-markets-get-help-through-innovation/ [Accessed July 11, 2021].

projects and co-creators of value. Indeed, construction companies are no longer conceiving solutions for their clients, but instead they are collaborating with them in this respect. In addition, they are connecting with one another to collaboratively identify problems and suggest reasonable solutions for their resolution. Previously, clients were neither involved in the conception of projects nor in the creation of value. Yet, nowadays, construction companies seem to be increasingly espousing collaborative governance models so they could serve their customers better. Through communication, construction companies are hoping to better understand the real needs and hopes of their customers.

> In construction, we cannot create a product and then consult with customers to adjust it based on their needs and preferences. This is simply unachievable in construction. What should be done is to immerse end-users in construction projects, from the very start till the very end. This is a precondition for the successful creation of smart cities (Mr. B.G.).

5.2.1.8. *Recognize end-users as both data sources and innovation creators*

As formerly noted, a smart project generally goes through various junctures and stages before completion, and for construction companies to be able to safeguard its smart character, they should ensure end-users are actively involved in the construction process.

> End-users would become partners to construction companies, acting as sources of data and co-creators of innovation (Mrs. S.P.).

Smart cities are humane cities – cities that are livable and dynamic. They are dynamic in the sense that their design allows for the effortless end-to-end flow of data. Indeed, data plays a crucial role in both the creation and administration of smart cities. Normally, data is gathered by public authorities and shared with everyone, especially young developers who could use it to create inventive things, practical solutions that may improve the standing of the city and boost its attractivity overall. Moreover, given the longevity of smart city projects, it would be a good idea for construction companies to create user manuals for housing and/or equipment in an effort to mitigate the risk of project-related data going down the drain over time.

5.2.1.9. *Co-create value by co-conceiving construction projects*

Data is today's new fuel oil – and effective data management practices constitute a solid foundation for the creation of smart cities. Regarding technology, it is a tool that is often used to manipulate and analyze data – so that plausible solutions could be figured out to resolve urban snags encountered.

> People are the source of data. And for data to be worthwhile, it should be exploited well; else it would turn out useless (Mr. B.G.).

Additionally, we should beware not to confuse smartness for the technology-intensive – as people remain at the heart of smart cities. Indeed, smart cities are people-oriented – and so, to be successful, they should preserve their human forms.

> By rendering data available for all, tech-savvy people would use it for innovative purposes (Mr. B.G.).

> For that reason, people are not only sources of data but value-creators too (Mr. B.G.).

Around the world, we come across various models of smart cities, ones that are mere technology-demonstrators (Songdo, South Korea) – and others that are instead focused on relationships and people (Copenhagen, Denmark). However, practical evidence has shown that technology-intensive cities were not as prosperous as expected; they failed to become smart and attract people and they remained inhabited and empty.

5.2.1.10. *Manage the stakeholders involved in the process and clear up roles and responsibilities*

Often, there is more than one major stakeholder in construction projects, especially large ones. A boost in the number of stakeholders (being often decision-makers, too), I would say, adds stress to the project and increases the project's complexity level. The ability of stakeholders to influence the project outcomes or execution approach would also affect the stakeholder complexity of the project. In addition to the number of stakeholders, the degree to which the project stakeholders agree or disagree on strategic matters relating to the project is likely to sway the project's complexity overall. A large construction project would typically involve numerous stakeholders. All the construction authorization agencies, environmental agencies, and labor and safety agencies (among others) have an interest in the project and could influence the execution plan of the project. The neighbors would have an interest in the architectural appeal, the noise and the purpose of the building. Based on the project, construction companies should conduct a stakeholder analysis or an assessment of a project's key participants to determine how the project would affect their problems and needs, as well as to identify their individual characteristics and interests, and find out what motivates them and what provokes them. In the GC BM, the GC should clearly define the roles and level of participation and decide whether there are conflicts of interest among different groups of stakeholders. The new BM is an abridged version of the old BM, involving lesser stakeholders who fulfill specific tasks and intervene at specific stages of the project lifecycle. In addition, by centralizing the decision-making process, we trust the project's complexing level will be pointedly reduced.

5.2.1.11. *Ensure apposite cueing and synchronization among construction actors*

Putting the conductor-contractor analogy aside (see Chapter 2), we herein avow that a GC somehow resembles a cook. The cooking process, similarly to the construction process, proceeds as follows: we cook, we taste, we rectify, we taste; if the final output tastes good, we deliver – otherwise, we rectify again, we taste and so on. Building, just like cooking, is an ongoing process. The only difference between the two is that in construction, adjustments are only possible at the very early stages of the process. By having an executive chef on premises, the time that would elapse before delivering the final output would be abridged – for the most part, because the latter would be practically flawless, requiring minor or even no tunings.

> In the restaurant business, cooking follows an agile methodology where tasks are broken down into small pieces and assigned to different team members. The final output would then be created once all pieces have been put together where each member would have contributed a piece to the final output. Could you imagine a big restaurant's kitchen being run without an executive chef where each team member works individually, not as part of a team? The situation would undoubtedly be chaotic (Mr. C.R.).

This comparison puts an emphasis on the importance of having a central operator in the construction industry who, like executive chefs, could synchronize construction works by cueing who does what and when.

5.2.1.12. *Rely on services of external partners to achieve specialized tasks (outsourcing)*

Today, for the efficient execution of smart constructions, big operators such as Bouygues feel the need to get involved, along with end-users and other key stakeholders, in the works carried out at the very early stages of construction processes, downstream of the value chain. By focusing on the Design-Build phases, they would be able to control the quality of works performed downstream and then ensure – in the Operate phase – that innovative value propositions would be eventually delivered to end-users positioned upstream of the chain.

> Though Bouygues, with the help of its subsidiaries, could cover the entire chain and handle construction works required all by itself, it frequently, for economic or other reasons, outsources part of the job to external entities (Mr. B.G.).

Nowadays, there are neither common guidelines nor plans *a propos* to how smart cities should be built. It is something that companies decide on according to the particularities of every project. Usually, project owners resort to subcontracting for the accomplishment of specific tasks, i.e. structure and digital infrastructure design, management, execution and promotion. *Per se*, subcontracting is good as subcontractors often bring expertise, cost efficiency and increased productivity to large projects. However, the problem lies in the fact that project owners do not hold trusting rapports with subcontractors. Generally, with no long-term commitments involved and assigned to realize very particular tasks, subcontractors tend to work disjointedly from one another, which results in incoherent, low-quality outputs.

> Today, there is a need for a prime contractor who could oversee the works carried out by subcontractors given that project owners are habitually incompetent in construction. The prime contractor, or the so-called GC, could fulfill such a task. Smart constructions are like puzzles where different pieces should be brought together to create a perfect final output (Mr. C.R.).

Hence, the GC's role would be to find the optimal combination of input during a given construction process from which would ensue a high-quality output. The GC would outline work standards to be adopted for the execution of construction works, manage staff and oversee their works, and handle disputes. Indeed, we assert, the GC BM, though based on a centralized management structure, is a collaborative solution that involves all construction actors, including both end-users and ICT companies. Like end-users, ICT companies (e.g. Schneider, Siemens, Orange, etc.) play a central role in the building of smart cities and could not be circumvented from the process, mainly because their contribution falls within the peripheral framework of the smart city.

> Habitually, public developers subcontract the digital infrastructural design of their projects to ICT companies. Through the technology implemented, the project built would end up being sustainable in the sense that resources could be allotted optimally, and operational expenditures end with the energy consumption cut in half (Mr. P.B.).

5.2.1.13. *Build a lengthy network of industry partners*

In France, large construction projects are often administered and executed by the so-called "*ensemblier*" – a group of companies (two or more) that partner together for the achievement of such projects. In the case of an "*ensemblier*", however, there are two or three decision-makers instead of one, meaning that a conflict of interest is still bound to happen at any point during long-lasting construction processes.

> The task of smart city management should be carried out by a single entity in partnership with other stakeholders: local communities to ensure there is social acceptability for the project, public authorities to ensure political acceptability and investors to ensure economic feasibility. In the realm of smart cities, value is often created in networks rather than individually, hence the importance of the GC role (Mr. P.B.).

Achieving a large construction project's goals takes a focused, well-organized person (or entity) who can engage with a committed team and gain the support of all stakeholders involved in the process. Building strong, trusting rapports with interested parties from the start could make the difference between project success and failure. Communication, being a key piece of stakeholder management efforts, should be preserved at all times. For construction companies, developing a communications plan that secures stakeholders' support is highly recommended. When it comes to GCs, they often have their own networks of subcontractors with whom they have been working for years and hold a rapport of trust and loyalty. The GC BM would hence allow for reduced turnover rates, abridged broken knowledge loops and increased stakeholders' engagement.

5.2.1.14. *Abridge logistics and procurement costs*

The construction industry is normally regarded as one of the oldest project-based industries. Indeed, the project-based managerial structure has remained intact for centuries in construction. Although it is one of the oldest industries, construction is still an unsustainable one in terms of the triple-bottom of sustainability, encompassing: economic, environmental – as well as social aspects. Focusing on economic means, construction is characterized by time- and cost-intensive production processes, which makes it prone to project risks and failure, in terms of time and budget.

In practice, this signifies that the performance of construction projects is usually low. Precisely, construction projects are regularly delayed and over budget. This is not only due to problems faced during project scheduling, but also during related processes, such as material procurement and material management. For many decades, methods such as the critical path method (CPM), the program evaluation and review technique (PERT) and Gantt charts have been applied in construction and have maintained their role for construction project planning. Though widely popular, these techniques do not pay sufficient respect to the complex planning environment in construction, and are solely suitable for the determination of time windows for project activities. Generally, construction companies – whether in France or elsewhere – use such tools to plan their projects; however, time and time again, they diverge from them, thus making them useless. The orchestration of time horizons in construction, we trust, should be made from within, through value chain planning. Specifically, the GC, by aptly synchronizing works, cueing intervention times and controlling the entire construction process (down and upstream the value chain), would guarantee the timely delivery of final outputs.

5.2.1.15. *Lower risk of premature technology obsolescence*

Today, the bulk of businesses, whether in construction or any other sector, are technology-intensive – and so, highly reliant on the adeptness of existent digital infrastructure at city or country level. Normally, the lifecycle of a smart project is five years upstream, five years downstream – 10 years in total, whereas that of technology is significantly shorter. This gap in life span between technology and construction projects, we presume, is likely to have a substantial impact on the economics of (smart) construction projects.

> Today, there is a need to shorten construction processes, mainly those specific to smart city projects, so they can mature after five years. This cannot be attained without the creation and adoption of innovative project management processes (Mr. B.G.).
>
> Delivery schemes will be more productive when digital technology (BIM) has been efficiently exploited in the design of construction projects, as this would enable the adoption of smarter manufacturing and construction techniques that speed up delivery, minimize disruption, and maximize efficiency (Mr. C.R.).

5.2.1.16. *Improve work quality and curb maintenance costs*

The construction sector is considered to be a basic industry that the development of the city (country) depends on to a great extent. In addition, the growth of construction companies is generally determined by their capabilities to innovate and espouse cost-efficient production processes. In construction, we proclaim, purchasing is a strategic function that could be used to control and reduce bottom line costs. Practically, this could be achieved by, among others: running value analyses (identification of hidden, unnecessary costs), making bulk purchases (specific for big operators), **setting tough quality control procedures**, dealing with suppliers located in the vicinity of construction sites (reduced logistics costs), **partnering with major suppliers** (sustainable construction materials).

Normally, quality management consists of two types of activities: quality assurance and quality control. Quality assurance relates to activities designed to ensure that quality is built into the process, whereas quality control is focused on activities designed to ensure that desired quality levels are actually achieved by the process. In construction, quality management is frequently achieved through quality control. Hence, we say that by regularly controlling the quality of works achieved throughout a construction process, cost reduction, productivity improvement, customer satisfaction, defect reduction and morale could be ensured. We finally note that quality control works best in settings where it is strongly supported by management, implemented by employee teams and there is a continual focus on process improvement that prevents errors from occurring. Under the GC BM, the GC would be in charge of controlling all works achieved throughout the lifecycle of a given smart construction project.

5.2.2. *Social*

We herein list and explicate the solutions suggested for the resolution of the formerly listed social (and other) problems.

5.2.2.1. *Promote collaborative urban governance schemes (social constructivism[34])*

In France, big market operators such as Bouygues are currently putting in a lot of effort to position themselves as leaders in the smart city field. Bouygues has recently been promoting a culture of integration by strengthening the rapport it holds with stakeholders situated at various levels of the construction value chain. Though the said company seems to be currently facing some issues controlling the works of its own subsidiaries, it is actually doing a great job promoting collaborative urban governance schemes for the building of smart cities. The collaborative approach being endorsed today would end up being beneficial for everyone, as all parties involved would profit from the exchange of knowledge and skills on the smart city topic.

> The city of tomorrow should be built collaboratively rather than individually, where every stakeholder contributes work and expertise to the equation, thus leading to the execution of high-quality construction works, and subsequently, the creation of an effective smart city (Mr. B.G.).
>
> The building of a smart city is mainly a question of urban governance. In opposition to Adam Smith's instructions, the sum of individual interests would not ensue in the fulfillment of the collective interest, unless individual works were properly controlled (Mr. V.B.).

34 Social constructivism. Like social constructionism, social constructivism states that people work together to construct artifacts.

Collaboration (referring to collective thinking and intelligence, i.e. smart governance), we proclaim, is hence indispensable for the success of large construction projects. Actually, project participants have nowadays realized that knowledge and information sharing is one of the key elements of a successful contractual relationship. As far as we are concerned, three factors can possibly boost the odds of construction actors working together, that is, enriched data sharing, improved quality, timely project completion and enhanced product and service quality.

> There is no clear guide today on the process of collaboration between stakeholders, therefore making it time and time again difficult to effectively interact and achieve common project goals within the bounds of cost, quality and time – thus the importance of your GC BM at this level (Mrs. A.T.).

5.2.2.2. *Partner with local communities for the conception and building of smart developments*

Today, local communities are making their voices heard when it comes to how they imagine the city of tomorrow should look and feel. In contrast to the past, they are currently more inclined to express themselves on topics – such as their preferred living locations, types of residences they would like to live in and others. In fact, just recently, local communities turned into key industry players who have a say in the design and conception of construction projects. Actually, they have become decision-makers who participate in the running of the city.

> For construction companies, it is crucial to integrate end-users in the construction process from the very early stages of the process (Mr. B.G.).

By collaborating with local communities, we believe that the cities of tomorrow will be managed more efficiently.

> End-users do not interfere in the technicalities of projects; they are incompetent in this area – which is normal. They will not take the pencil and start drawing the city, but instead, they would contribute ideas; ideas that would help shape the look of smart cities (Mr. B.G.).

Therefore, construction companies should listen to customers and jot down their comments. Ultimately, they are the ones who would potentially purchase the final product – and for them to be inclined to do so, they should be satisfied with the product in the first place. Communication channels between construction companies and local communities should remain open throughout the entire project process.

> The partnership between the two is an ongoing process, a long-term relationship that can only be broken upon mutual agreement (Mr. B.G.).

In the realm of smart cities, it is the tiniest details that count the most: would I build a bakery or a library on the street corner? Though the question may sound silly, such an information makes the difference in the success or failure of a construction project.

> By partnering with local communities, construction companies would be able to build not only walls and roads, but the spirit and soul of cities (Mr. B.G.).

5.2.2.3. *Rely on collective thinking and intelligence to conceive and execute smart developments*

The problems encountered today at the level of the French construction industry, we proclaim, could be solved by encouraging both collective planning and collective thinking.

> The solution lies in centrally planned collective housing to be ensured via resourceful public–private partnership schemes, where the public sector governs and the private sector builds. Besides, given that public authorities are often incompetent in construction, the governance of large projects would end up being delegated to a maven private entity (Mr. D.B.).

Consequently, the private sector would be in charge of running the projects, ensuring both governance and execution of operations.

> In France, one in four homes are centrally planned and follow a style of collective housing. Social housing is the type of housing that honors its social heritage the most (Mr. V.B.).

Within this setting, the GC role would be invaluable in promoting this type of centrally planned, social housing construction.

5.2.2.4. *Create value in networks – profit sharing logic*

As previously stated, the governance of smart cities is better ensured by a GC – not alone, but in partnership with end-users. In our eyes, smart cities are primarily a question of smart governance and smart people – and it is their smart collaboration that would safeguard the smart character of cities.

> End-users should assist GCs in the management of smart cities. Together, they form an ensemblier – who would run smart projects based on a DBO (Design-Build-Operate) model (Mr. D.B.).

We conclude by pointing out that the success of smart projects relies on the ability of construction companies to build and manage them following a people-oriented economic model, and based on a bottom-up managerial approach (thus, the solution suggested next!).

5.2.2.5. *Administer smart developments following a bottom-up managerial approach*

Créa'City[35] is a good example of a smart initiative (program) that encourages partnership between local communities and construction companies – and supports personal (innovative) initiatives. Créa'City is a kind of brainstorming session where different people sit together to discuss the fundamentals of sustainable life (reacting to questions such as, what should the city of tomorrow should look like?). Undeniably, the involvement of end-users in the very early stages of large construction project processes is a prerequisite for their success.

> Often, people are right and construction companies are wrong; the assumptions companies tend to make about people's needs are time and time again over erroneous. We converse, we make adjustments, we fix things, and then we converse again. Smart projects are subject to constant adjustments until the point construction works have been initiated (Mr. B.G.).

> It is not up to big operators, such as Bouygues, to decide what people's real needs and aspirations are, as citizens are the only ones who are entitled to speak for themselves (Mr. E.N.).

5.2.2.6. *Set up effective stakeholder management strategies*

In the realm of smart cities, construction companies must work on enhancing cross-functional collaboration across the value chain and building long-lasting rapport with stakeholders. As formerly evoked, construction projects have a lot of fuss and they could easily get off track if the various stakeholders are not working together efficiently. The importance of collaboration for more effective construction projects is increasingly recognized across the construction industry. Indeed, improved collaboration is expected to reduce risk on construction projects and boost the sense of engagement and responsibility among stakeholders. Big operators often

35 Available at: https://www.creacity.be/.

use tools such as the BIM[36] to improve multi-stakeholder, real-time communication. When it comes to France, despite the widespread agreement about the value of increased collaboration, the country's construction industry is currently still struggling to be collaborative and create sustainable value networks.

In order to run a successful construction project, it would be central for a construction company to address the needs of the project's stakeholders, effectively predicting how the project would affect them and how they would affect the project. Ineffective stakeholder management practices could eventually result in dissatisfaction with the final output and negative impacts on the project's budget and schedule. During the development stages of a construction project, we proclaim, a clear and detailed stakeholder management strategy should be established. While it may be hard to satisfy all stakeholders on every project, effective stakeholder management must aim to satisfy as many as possible, which would often involve strategic prioritization of different stakeholders needs. From our viewpoint, an effective stakeholder management strategy could take the following form: inform–consult–involve–collaborate. The first category (inform) includes those stakeholders who require minimal effort. Stakeholders with lower power but higher legitimacy need to be kept informed about decisions taken that may affect them directly. The second category (consult) covers stakeholders who require more than just being informed about the project. Those who have higher power but lower legitimacy should be kept onboard and consulted, in order to seek their opinions and input for key decisions that directly or indirectly affect them (usually, companies are unlikely to change their strategies based on the results of such consultations, but may adjust their tactics to maintain high levels of commitments). Last but not least, the third category (involves) comprises stakeholders with high power levels, even if they have low legitimacy. They need to be involved in all project activities according to their interest, since they have the power to make decisions that impact on the project. Finally, the fourth category (collaborate or partner) involves primary stakeholders who have high levels of legitimacy and power to affect the success of the project – and as such, they should be treated as partners to increase their engagement and commitment. This is often achieved (whenever necessary) by revising and tailoring project strategies, objectives and outcomes[37].

---

36 BIM stands for Building Information Modeling. BIM refers to a collaborative method of working which is based on the generation and exchange of data and information between the various project parties. Based on this information, the full cycle of a building, from conception to completion, can be managed.

37 Stakeholder management for building design and construction. Available at: https://www.designingbuildings.co.uk/wiki/Stakeholder_management_for_building_design_and_construction#Stakeholder_management [Accessed July 11, 2021].

5.2.2.7. *Enhance end users' well-being and quality of life*

In France, large construction projects often fail to sell because they greatly diverge from people's exact needs and aspirations. We, as construction companies, promise end-users one thing at the start and then deliver a different thing at the end. We simply fail to keep our promises.

> It would be beneficial to delegate the management of smart constructions to a maven GC, who may ensure that the promises made downstream the chain would hold true upstream (Mr. B.G.).

Precisely, the GC would have a certain control over the value chain and ensure that the final product meets people's real aspirations and hopes.

> A smart construction is part of a smart city – and so, construction companies would eventually fail to create smart cities unless they ensure their constructions are smart; a smart city is a chain of smart projects (Mr. B.G.).

It is a solution that construction companies offer, value propositions that go beyond the simple act of selling mere structures. Indeed, smart cities must mirror people's real needs and territories' embedded values.

> To enhance the attractivity of cities, construction companies must step aside and its inhabitants must take over. Construction companies must connect with citizens by raising their awareness about the fact that it is the little actions they undertake that will influence the future look and shape of their cities (Mr. E.N.).

We accept that smart cities cannot be built without trial and error: we try, we fall, we get up and then we try again. Hence, by collaborating with other stakeholders, we will be able to identify the real problems that are holding our cities back and come up with solutions that could allow us to move ahead.

> Today, there is a shift from the technical to the methodical. Indeed, today, people are less concerned with the performance of the technical services provided; what is instrumental for them is rather the performance of the social service that accompanies the technical (Mr. C.R.).

This is practically accurate as construction companies are now not only delivering technical performance, but also social performance. Social performance, we say, relates to the well-being and quality of life of people. In addition, people's well-being is not preserved through the provision of, for instance, well-connected accommodation or cutting-edge heating systems, but lodgings that are located in pleasant sports, in nice neighborhoods, alongside charming street corner bistros. Hence, we acknowledge that the performance of a smart project is both financial and social.

> People would be willing to pay more to live in an enjoyable neighborhood (Mr. B.G.).

Normally, it takes time to build smart cities. They are often built successively, piece after piece, year-after-year. Indeed, over time, the raw idea that we had would materialize to convert into something tangible. Moreover, smart cities are collaborative systems where stakeholders join forces to put together the different pieces of the puzzle. If properly conceived and built, smart cities would be beneficial for all.

> Developers should be ready to make some sacrifices, by say, allotting a 500 m^2 area (equivalent to one million euros) for the building of green spaces, gyms, yoga spots, and kids zones (among others) – to boost the well-being of local communities. Even if such offerings are not always sufficiently payable, they create sympathy and enhance the saleability of the final product. Enhancing social well-being implies adding value to the property, thus making it understandably more expensive (Mr. B.G.).

Moving on, in the wider property, the key focus of sustainability has been and still is to a greater extent, on environmental impacts (e.g. global warming and resource depletion). These are undeniably crucial, but in our enthusiasm to tackle one problem, we ignored another equally important problem, that is, the impact of buildings on people and especially, the impacts of interior spaces where they spend most of their time. We know that sustainable buildings ought to support and help people live better, healthier and more productive lives, but this metric has been marginalized or seen as irrelevant in some cases up until now. People live indoors most of time. Buildings directly influence occupants' health and productivity. The effect of indoor environmental quality in buildings on occupants' health, well-being and productivity is an important topic in occupational health and public health. To my knowledge, buildings, which are facilitated by green features, provide a clean and more comfortable environment for their dwellers. Also, better ventilation, higher indoor air quality, environmentally friendly materials and optimized lighting are green building features that may positively influence people's well-being and health. Hence, smart constructions – as part of smart cities – are expected to bring well-being to people today, while preparing the future for all.

5.2.3. *Environment*

We herein list and explicate the solutions suggested for the resolution of the formerly listed environmental (and other) problems.

5.2.3.1. *Organize hearing sessions for all people affected by the project*

A smart project is often lengthy and may take years to complete, which signifies that its impact on the environment – and so, on local communities living in its vicinity – could be sizeable, especially if no counteractive measures were clearly defined at start. In fact, whenever a company is handling a smart construction project, it has to think about the budding environmental impacts that could ensue from it. In addition, local communities living in the vicinity of the project should be kept informed about the works to be undertaken, as well as the benefits and costs that could result from it as those may influence them, their lives and well-being.

> Construction companies must organize hearing sessions for all people who may be directly or indirectly affected by the project. Moreover, in some specific cases, people should be entitled to financial compensation for the damages experienced during the course of the project (Mr. B.G.).

5.2.3.2. *Favor ecofriendly constructions*

From our viewpoint, eco-friendly construction begins with a smart design. By thinking strategically and opting for the right materials and technology early, construction companies may contribute to saving the environment while cutting project costs. Smart design means making the best use of space possible. It is a motive for developers to start thinking ingeniously about space – and the future of infrastructure. Something as simple as the placement of windows could make a great deal of difference to the atmosphere of a room. Innovative technology for construction companies assists in identifying the most strategic window positions to take advantage of sunlight and natural breezes. Insulation is another thing to consider when building eco-friendly properties. On a separate note, we avow that the life expectancy in architecture and construction is currently growing in importance, and that sustainable construction materials (e.g. brick, stone and concrete) have a longer lifetime and are more eco-friendly during the demolition and disposal phase. Extensions, renovations and refurbishments stretch the life of a building, specifically when it is constructed using sustainable materials.

5.2.3.3. *Use additive or other technologies and systems to boost energy efficiency*

One of the early signs that additive manufacturing had made it in architecture was the shift from its use by designers, solely as a tool for prototyping, to its use in fabricating finished objects. Although slower than many mass-production processes, the increased design control and scalable customization afforded by the new technology through processes – such as 3D printing – has propelled its rapid growth within many industries[38]. Thanks to newfound capabilities, particularly relevant to architecture, additive manufacturing is taking another step forward: printing buildings *in situ*.

This capacity is possible mainly because of advances in concrete fabrication technologies. The Chinese construction company, WinSun, for example, made headlines when it constructed 10 single-room houses in Shanghai from 3D-printed modules in less than 24 hours. Although these were built from components that had been fabricated offsite, the company is now planning to produce entire structures – including bridges and tall office buildings – onsite. WinSun's material feedstock is a combination of cement, steel, glass fiber and recycled construction debris. The use of waste materials is a clever strategy to diminish the carbon footprint of concrete – and the printing process ensures that construction workers are less likely to be exposed to potentially hazardous ingredients.

> Nearly one-third of the global population currently lives in earthen buildings. The clear environmental and economic benefits of this construction method has inspired the recent development of novel strategies that combine onsite earthen construction with additive manufacturing techniques (a passage extracted from Earth Architecture, Princeton Architectural Press 2009) (Ronald Rael).

Moving forward, we affirm that construction companies should start integrating the environment into their construction processes. They should also provide end-users with service solutions, a set of smart things that could be implemented at city level in an attempt to boost its attractivity and livability. In reality, this, we say, could be achieved by focusing on two areas: (1) the improvement of waste management processes and (2) the extensive use of RES[39]. Monaco is a good example of a smart, self-sufficient city-state that has very efficient waste

---

38 Using Additive Manufacturing to Build With Materials Sourced From the Jobsite. Available at: https://www.architectmagazine.com/technology/using-additive-manufacturing-to-build-with-materials-sourced-from-the-jobsite_o [Accessed July 11, 2021].

39 Technical terms and acronyms in construction available at: https://www.allacronyms.com/BM/construction.

management systems. In Monaco, rather than data, waste is today's fuel oil. Indeed, through waste-to-energy technology, Monaco now has the capacity to process 80,000 tonnes of waste per annum with energy recovery.

Concerning RES in particular, these can be used to promote sustainable constructions. For instance, a geothermal system harnesses temperatures below ground to warm residential homes or businesses. Pipes, connected to a heat pump, can either warm or cool buildings. Solar power converts the sun's radiation into energy. Installing solar panels may be expensive today, but in the long run saves a lot of money and energy consumption. The carbon footprint of these technology-led developments was reported to be significantly lower than traditional constructions, as well as being cheaper to run, and producing less waste[40].

In conclusion, we highlight that the development and deployment of digital technologies and processes – value chain digitalization – is central to the required transformation of the construction industry. Innovations of this kind, we proclaim, would unquestionably enable new functionalities along the entire value chain, from the early design phase to the very end of an asset's lifecycle at the demolition phase. According to a recent study[41], full-scale digitalization (the use of big data and analytics, as well as new simulation and virtual reality methods) in nonresidential constructions would be, within 10 years, capable of producing annual global cost and resource savings of nearly 1 trillion dollars on engineering and construction. Indeed, by exploiting mobile connectivity and augmented reality, construction companies would be entitled to progress the performance of their production processes by engaging in real-time communication and providing stakeholders with additional on-site data.

5.2.3.4. *Use of innovative, ecofriendly, cost-effective construction materials*

Just as the definition declares, sustainability is meant to make something last. The main purpose of using sustainable building materials is to construct homes that last – and longer-lasting homes mean lower maintenance costs and less upkeep. A cost reduction and extended life span are two remarkable benefits that could draw construction companies' attention. Apart from the economic benefits, other, nonfinancial incentives may also push construction companies to use sustainable

40 Scientists are taking concrete steps towards reducing cement's massive carbon footprint. Available at: https://qz.com/1748561/%E2%80%A8reducing-cements-carbon-footprint-is-critical-to-climate-fight/ [Accessed July 11, 2021].

41 Shaping the future of construction: A breakthrough in mindset and technology. Available at: (http://www3.weforum.org/docs/WEF_Shaping_the_Future_of_Construction_full_report__.pdf) [Accessed July 11, 2021].

construction materials – explicitly: decreased carbon footprint, increased home resale value, shorter on-market periods (among others).

> To my knowledge – and based on previous conversations with peers, "a home that employs third-party certifications for sustainability may, on average sell for 10-15% more than a similar non-certified home".

5.2.3.5. *Encourage smart and ecofriendly initiatives (urban-village)*

When creating a smart city, environmental issues should be prioritized – and be seriously addressed (e.g. energy efficiency, carbon footprint, building energy performance, etc.). In Paris, the building of eco-districts is now on the go. Though these constructions are not smart *per se*, they are sustainable and eco-friendly. In the realm of smart cities, construction companies' priorities must shift from being purely economic to environmental – and social (although not necessarily in that order!). Indeed, in their projects, they must ensure some sort of evenness between all three dimensions. Today, if we ask a client, where do you prefer to live or what amenities do you seek most, their answers would mostly relate to the environment. This partially explains the growing interest among companies for the environment – and why the spectrum of construction projects is lengthening beyond the structures built *per se*, to cover their surroundings – and sometimes the entire neighborhoods in which they are sited.

Generally, people choose where to live based on two factors: the atmosphere of the area (livability and mobility) and the living environment (cleanliness, open spaces, natural sights, greenery, etc.). This is, we proclaim, what smart cities are all about – the summary of its components – including the environment.

> Nowadays, construction companies are building milieus rather than mere constructions; places where people would be willing to live and social collective welfare could be upheld. Eco-friendly, smart initiatives are much needed today in France – something like the Chinese urban-village concept; they should be encouraged, endorsed, or even sponsored (Mr. B.G.).
>
> Butte-aux-Cailles is a great example of a French urban-village (Mr. E.N.).

5.2.4. *Technical*

We herein list and explicate the solutions suggested for the resolution of the formerly listed technical (and other) problems.

5.2.4.1. *Assign central operator to oversee execution works*

Today, we do not actually have to reinvent the wheel, as many available tools could be used to restore the construction industry's situation. Generally, the management of big projects is a complex affair, thus the importance of relying on specialized technologies, such as BIM to ensure that construction projects are managed efficiently, all stakeholders have access to the same data, and work progress is trailed on a regular basis. In addition, there is a need to centralize decision-making, where authority and control are placed in the hands of a maven entity that could push construction works forward and safeguard the quality of the product throughout the entire construction process. This is in fact what the suggested GC BM is all about. Under the said model, we clarify; the GC would be in charge of all construction works, from initiation to closure. They would also take on all of the risks associated with the project. As for the project owner, they could partake in the very early programming stages (e.g. deadlines, validations, documentations, etc.) of the project – and then withdraw relaying to the GC who would take over and act as a delegated project owner

> The suggested GC BM, slightly resembling a concession model[42], may indeed serve as a solution for the various challenges encountered at the industry level today. The project owner would subcontract a GC who has sufficient knowledge and expertise in construction for the handling and managing of smart constructions. Under the GC BM, the centralized entity in charge of the entire construction process – that is, the GC – would be ultimately held responsible for the success or failure of the project (Mrs. A.T.).

> A conductor or an integrator role is definitely necessary in the construction sphere today. It should be pointed out that, although such a role could be efficiently played by a GC, it could also be apportioned to several entities rather than just one. The main concern at this level lies in the fact that large projects are the proprietorship of public entities who, routinely, through passive bidding processes, commission them to the cheapest, often incompetent bidders. This puts a question mark on the ability of project owners to delegate their roles to the right entities (Mrs. J.N.).

---

42 The concession model involves a designated space in a department store that operates somewhat autonomously – with its own signage, walls, furniture and displays; it provides its own staff and has more control over distribution by moving away from the wholesale model and selling products directly to consumers.

5.2.4.2. *Rely on organizational change to drive industry growth*

Given that large construction projects are often state-controlled, we believe that the GC (being the delegated project owner) would have a somewhat limited supervisory role, rather than a decision-maker or conductor role.

> Large construction projects are often lengthy and associated with a bunch of operational issues. And so, if a GC or any other competent entity comes to replace the incompetent project owner, this would unquestionably mark the debut of a period where the role of public authorities in the construction sector has started to fade. The GC would play a crucial role by bringing all stakeholders together – and the change they would bring must focus on the downstream, that is, the very early stages of a construction process. The rest would automatically take care of itself (Mrs. A.F.H.).

In our eyes, the GC BM we are proposing is comparable to the Design-Build delivery system, where the project owner contracts construction works to only one entity. Under the Design-Build delivery system, the owner only communicates with one entity, whose responsibilities and liabilities are clearly defined. Cost and time savings could be realized as stakeholders would work in tandem, synchronically.

Likewise, we trust the GC BM to overlap slightly with another well-known construction project delivery system, which is based on the engagement of a CM, who provides professional construction management services (including construction costs, scheduling safety procedures and others) at the early stages of construction processes. Under the GC BM, the GC would be handling such tasks.

Moreover, resemblances could be dotted with the PM delivery system, where the owner hires an independent PM to manage the entire project on his behalf. The underlying assumption at this point is the outsourcing of the responsibility for the entire project to a separate entity on behalf of the owner. This might be the case if in-house experience is insufficient to undertake a construction project and impermanent professionals are required to arrange and control the planning, design and construction process. We lastly shed light on the fact that the GC BM is also comparable to the build–own–operate–transfer system in the sense that, in both systems, the owner solely contracts with a single turnkey contractor. Under the GC BM, based on the project scope and the objectives assigned to it, the GC is expected to play several roles, that of a CM, a PM, as well as a turnkey contractor (refer to Chapter 1 for more details in this regard) (see Figure 5.1 for a simple illustration of the GC BM).

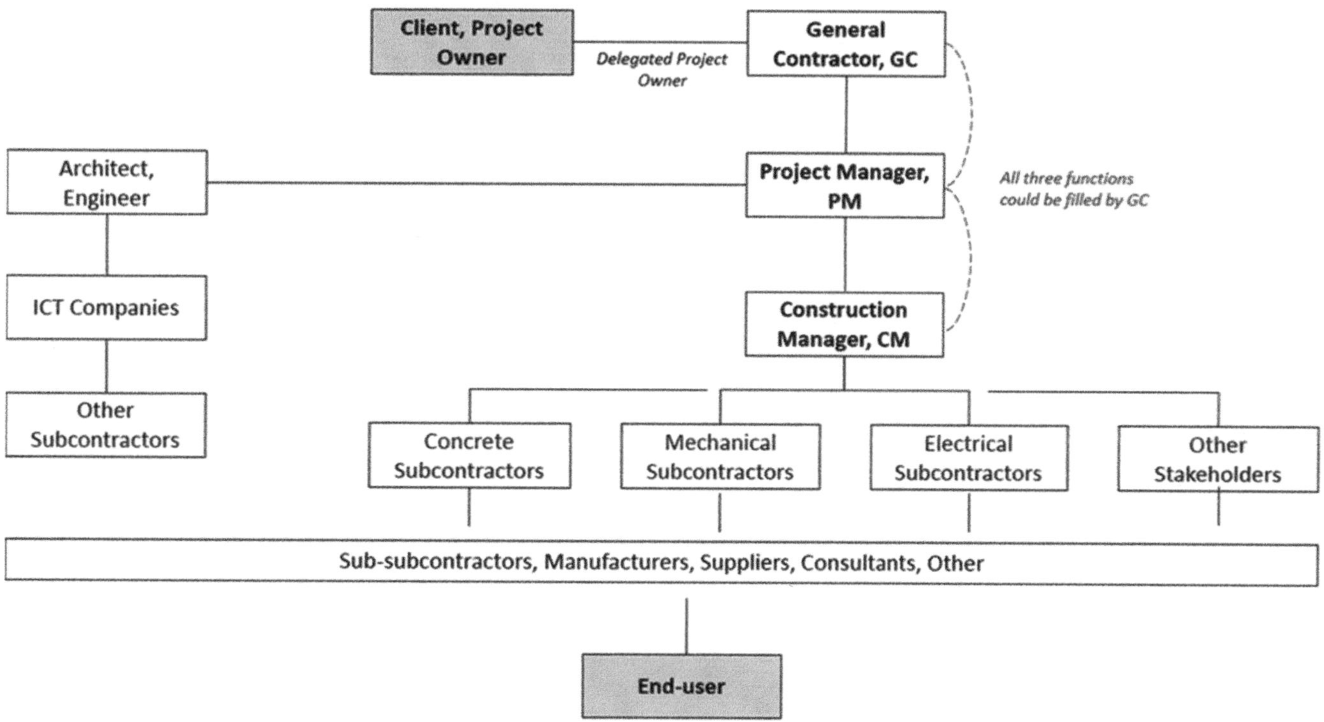

Figure 5.1. *Simple graphical illustration of the GC BM (source: created by the author). For a color version of this figure, see www.iste.co.uk/karam/general.zip*

> What the construction industry needs today is an organizational change, a new project management model, that construction companies could opt for to competently oversee and manage large construction projects. Under the GC BM, the GC would ensure that the delivery of the project is on time and within budget; decide the workflow (what to do, when and where, and by whom?); synchronize construction works; and signal opportune intervention moments for stakeholders (Mrs. A.T.).

Perceiving things from a slightly different angle:

> Innovation in construction should now focus on the resourceful design and management of ground floors. Apposite urban governance requires the ground floors of buildings to work symbiotically with the surrounding sidewalks and public spaces. Together, they provide a continuous network of pathways and experiences that are active, safe, comfortable and engaging. Innovation often ensues from the little things we do, the incremental improvements we bring to our business models, to our constructions (Mr. B.G.).

5.2.4.3. *Set cohesive plan for the execution and tracking of construction works*

A smart city is a blend of different components that are linked together, thus forming a coherent system where people can work, live and prosper. The cities we live in today were built hundreds of years ago and the changes we bring to them are often done gradually. What we are doing today is adapting our cities so that they can become smarter. Moving on, it is worth mentioning that smart constructions are complex; they go through various stages with a lot of intervenors stepping in and out at each stage. Deciphering the boundaries of smart cities and matching our offerings with customers' needs are surely important, yet what is really needed today is a cohesive plan for the execution and tracking of large construction works.

> There is definitely a need today for a managerial innovation that would help construction companies to execute large construction projects skillfully – and create healthy ties with stakeholders operating down and upstream the value chain (Mr. D.B.).

> Whenever we think about smart cities, it is the governance component that we should focus on and improve (Mr. B.G.).

This does not relate to the technicalities of projects, but rather to the way projects are being executed and managed. The central idea here revolves around the ability of construction companies to efficiently manage their day-to-day operations (e.g.

define clear objectives and work plans, set precise timelines and budgets for the execution of projects and others).

> Under the GC BM, the GC would be consigned by the project owner to ensure the smooth running of large construction projects (Mr. D.B.).

5.2.4.4. *Better innovation management practices*

The construction industry in France (and around the world) has long been urged to innovate, but innovation has been elusive because of the inherent social and organizational complexity of construction. Focusing on the practice of innovating requires exploring the enabling capability of solutions for practitioners to establish novel ways of doing things for improvement, referred to as – innovative capability. BIM has been promoted as an enabler of innovation in construction design because of its data management capabilities and the opportunities for interdisciplinary work based on them. Today, the construction industry seems to be underperforming and failing to deliver optimum value, and the lack of communication and coordination among various stakeholders has been seen as a major reason for this. Consequently, the need for innovation to improve communication and coordination has been a recurrent theme. Nevertheless, the construction industry and its projects exhibit social and organizational complexity, which makes innovation difficult to define, implement and use. This suggests that practice is the locus of innovation in construction. Therefore, developing an insight into the practice of innovating is necessary in order to produce knowledge on innovation that has a practical application. Indeed, what France-based construction companies must do today to improve their innovation management practices and succeed their smart city projects is to find a median-solution, an alternative to two extreme approaches, technology-centered and human-centered.

5.2.4.5. *Set efficient budgeting systems and flexible timelines*

A budget is a statement of the amount of money that is available to spend over a period of time, or on a specific thing. It may comprise an outline plan for how that money would be spent and a breakdown of the items it would be spent on. Overall, budgets for construction projects help determine what is affordable – and should be set as early as possible. It is central for a project budget to be realistic and based on evidence. A project budget could be established through the assessment of projected income and expenses throughout the lifecycle of the project, in comparison with similar projects, and pre-design analysis of requirements (among others). Usually, the project budget is set by the project owner and is distinct from cost plans prepared by a construction company, which are likely to solely focus on construction costs. In large construction projects, it is common that the budget diverges briefly over time and it is for this reason that careful cost control is important (this task is handled by the GC, under the GC BM). It is also important that the project owner makes it clear what costs should be monitored by the construction company and what should remain within his control. Architects often design in a vacuum. Cost and budget are key constraints that should always be identified and considered when designing a construction project, even when the provision of cost advice is expressly excluded from the architect's obligations. Like construction budgets, construction timelines should be realistic and flexible, accounting for unforeseen events that may occur during the lifecycle of a project and delay its execution. A proper construction timeline is one that allows adequate time for the delivery of supplies, as well as time for unfortunate and unforeseen hassles like order mix-ups, construction snags or inclement weather – just to cite a few. In Anglo-Saxon countries, successful projects often start with a maven GC – an experienced GC who has ample *savoir-faire* in the general execution of construction projects brings so much to the table when it comes to thorough construction management.

> Even though hiring an expert GC can sometimes be costly, the connections and resources they are likely to bring with them would be beneficial for the client and the project in the long run (Mr. B.G.).

5.2.4.6. *Assign specialized entity to handle after-sales services and maintenance works*

In construction, maintenance is the process of ensuring that buildings and other assets retain a good appearance and operate at optimum efficiency. Inadequate maintenance could result in decay, degradation and reduced performance and could affect heath and threaten the safety of users, occupants and others in the vicinity. Depending on its design, quality of materials and workmanship, function and location, buildings deteriorate at different rates and require different levels of attention. Although no building will ever be maintenance-free, the quality of the design and workmanship could sustain the final output by minimizing the level of maintenance it requires. Under the GC BM, the role of the GC would extend to post delivery of the final product, to handling after-sales and maintenance services. This is a service solution – in response to inadequate maintenance and responsibility diffusion problems encountered in projects built and delivered under the old construction industry's BM – by which the GC guarantees assets would retain a good appearance and operate at optimum efficiency over time.

Table 5.2 summarizes suggested solutions for problems encountered in the French context.

CONSTRUCTION STAGE	*BIM-PROCESS*	*SOLUTIONS*			
		ECONOMIC	*ENVIRONMENTAL*	*SOCIAL*	*TECHNICAL*
INITIATION & CONCEPTION	*PROGRAMMING*	*– Espouse a centralized decision-making process – Ensure final product has a use-value and that utility may be extracted from it – Adjust offerings to exact needs and requirements of end-users*	*– Favor eco-friendly constructions*	*– Promote collaborative urban governance schemes (social constructivism) – Partner with local communities for the conception and building of smart developments – Administer smart developments following a bottom-up managerial approach*	*– Assign a central operator to oversee execution works – Rely on organizational change to drive industry growth – Set efficient budgeting systems and flexible timelines*
	CONCEPTUAL DESIGN	*– Ensure involvement of end-users in all stages of the construction process*	*– Organize hearing sessions for all people affected by the project*		
PLANNING	*DETAILED DESIGN*	*– Recognize end-users as both data sources and innovation creators – Co-create value by co-conceiving construction projects – Allow for customizability (portfolio diversification)*		*– Rely on collective thinking and intelligence to conceive and execute smart developments*	
	ANALYSIS	*– Identify and target underserved market segments*	*– Encourage smart and eco-friendly initiatives (urban–village)*		
	DOCUMENTATION				*– Set a cohesive plan for the execution and tracking of construction works*

LAUNCH & EXECUTION	*FABRICATION*	*– Rely on the services of external partners to achieve specialized tasks (outsourcing)* *– Manage stakeholders involved in the process and clear up roles and responsibilities* *– Ensure apposite cueing and synchronization among construction actors* *– Build a lengthy network of industry partners*	*– Use of innovative, eco-friendly, cost-effective construction materials* *– Use additive or other technologies and systems to boost energy efficiency*	*– Set up effective stakeholder management strategies*	
	CONSTRUCTION 4D/5D	*– Lower risk of premature technology obsolescence*		*– Create value in networks: profit sharing logic*	*– Better innovation management practices*
	LOGISTICS	*– Abridge logistics and procurement costs* *– Improve work quality and curb maintenance costs*			
MONITORING & CONTROL	*OPERATION & MAINTENANCE*	*– Improve attractivity and absorption capacity of city*			
CLOSURE	*OPERATION & MAINTENANCE*			*– Enhance end-users' well-being and quality of life*	*– Assign specialized entity to handle after-sales services and maintenance works*
RESTORE, RENOVATE, REMODEL	*DEMOLITION & RENOVATION*				

Table 5.2. *Key solutions in construction (France) (source: created by the author)*

5.3. The general contractor business model

Frankly, it is the existence of various problems in construction, which remain unsolved to date, that encouraged me to think of a GC BM for the building of smart cities in the first place. Using the GC BM, problems identified and solutions suggested for their resolution were mapped across different dimensions of construction projects. For this, we used the TLBMC, being part of our GC BM, to which we added a fourth layer: technical (stratum 4). This addition was compulsory, as we were not able to fit all of the problems identified into the envelopes of existing layers: economic, social or environmental. From our viewpoint, some problems are rather technical, relating to the technicalities of projects, with the ability to sway the productivity of construction projects, including passive bidding processes and legal and institutional barriers. Therefore, by adding an extra layer, we have widened the spectrum of the TLBMC. In addition, as you will have noticed by now, some of the identified problems were labeled off-balance. Those are problems that we, as construction companies, have zero control of, yet could pointedly sway our planning and project management capabilities. As exemplified in Chapter 4, some of those problems founded the boundaries of our GC BM.

The GC BM, as its name suggests, revolves around a central operator – a general contractor, consigned to manage and oversee groundworks in large and complex construction projects. They oversee the whole construction process, from initiation to closure. As a delegated or acting project owner, they replace the project owner throughout the entire construction project lifecycle. Endowed with necessary power and authority, the GC, a maven decision-maker, is entrusted to make the right decisions at the right times. They assume all of the risks associated with the project and are accountable for its success or failure. Ensuring interoperability between stakeholders involved in the process is among their key tasks. For the most part, their role consists of synchronizing construction works and cueing intervention moments. As discussed throughout this book, the GC resembles a music conductor (and a cook), whose primary obligation is to create coherent arrangements, allowing the production of quality outputs. The GC's role is managerial by nature. They ensure control over the quality of construction works and keep communication channels open between all stakeholders involved in the construction process.

Putting the GC role aside, under the GC BM, end-users become part of construction processes, as construction actors who partake in the process from the very early stages of initiation until the very late stages of closure and delivery. Indeed, they act as co-creators of value and innovation. Overall, the GC BM is a collaborative governance model where all stakeholders work together, each contributing something to the final output, among others: architects handle drawings, ICT companies work on the digital infrastructure design of the project and construction staff ensure construction works are delivered on time. It is through collaborative work and effort that smart cities are constructed.

As previously evoked, the GC BM is made up of six connected strata (see Chapter 4 for further insights into this). The pivotal one is stratum 1: the construction stage. This is factual in the sense that construction projects are stage-based and that problems faced in construction are likely to vary on a stage basis. Moving on, given that innovation is technology-enabled, we chose to add stratum 2 to our GC BM: BIM process. BIM benefits are numerous, including four main efficiency factors: improved collaboration, reduced costs, time saving and increased sustainability[43]. Regardless of the practicality of BIM technology, the latter remains misused by France-based construction companies today. Stratum 2 has been integrated into the model in the same way that it was imbedded into stratum 1. This enabled us to split construction stages into smaller pieces so that the tracking and management of construction works could be performed at a more micro level. Construction actors denote stratum 3 of our GC BM. As recurrently evoked, there are numerous stakeholders involved in a construction process, stepping in and out of the value chain during the life span of a construction project. They are likely to change each time we move through the stages of a project. Thus, by manipulating stratum 1, and accordingly stratum 2, we could have a visual of who does what (stratum 3) and at which stage (strata 1 and 2).

Regarding stratum 4, it represents the different dimensions of a construction project. Those are symbolized by the four TLBMC layers: economic, social, environmental and technical. Besides, given that our BM is catered for the building of smart cities, it seems obvious that smart city components should be a part of it, in stratum 5. Moving on, instead of directly linking the TLBMC layers to their corresponding building blocks, we have included a control variable in between the two: smart city components. This stratum allows companies to track construction activities while ensuring that the different constituents of a smart city are accounted for in their projects. Finally, stratum 6 represents the TLBMC building blocks. By varying inputs at level of stratum 4, the ensuing output at the level of

43 Four BIM Cost Savings and Efficiency Factor. Available at: https://www.kreo.net/blog/bim-cost-savings-and-efficiency-factors [Accessed July 11, 2021].

strata 5 and 6 would automatically change. In our GC BM, there are three strata that could be used for simulative purposes: one, two and four. We note at this point that a business model simulator, scripted in Angular 2+, was created for the running of real-time simulations. Constructed using the funnel technique – going from the most general to the most specific, our GC BM is original in the fact that it is a BM as much as it is a piloting tool – precisely, a dual-use BM. Strata 1 through 6 form the piloting tool, whereas strata 4 through 6 form the business model. Finally, it is worth noting that our GC BM is both solution-based and multi-sided: solution-based because we use it to map problems faced at each stage of the construction process, and also categorize them based on their types, and multi-sided because the GC plays the role of a connector, ensuring that the works performed throughout the lifecycle of the project roll harmoniously and synchronically. They also act as an intermediary who transmits data along the value chain, end-to-end, from project owners, downstream – all the way up to end-users, upstream.

5.3.1. *Breakdown of problems*

The breakdown of problems identified in earlier sections was done using the GC BM, minus two strata: stratum 2 (BIM process) and stratum 5 (smart city components). Indeed, we use a less extensive version of the GC BM, a four- rather than six-stratum BM, for the reporting of problems faced in construction today. We do so for one forthright reason, that is, under current circumstances, BIM technology and smart city components are not integral parts of the prevalent construction industry's value chain in France (see Table 5.3 for a sneak peek into the breakdown of problems in construction across the TLBMC's layers and building blocks[44]). In the next section, we present a breakdown of problems faced in construction today and explain existing dependencies among them. The problems identified are dispensed across the various building blocks of the GC BM – there are a total of 34. For the explanation of dependencies, we select a few problems and explain them using texts and graphs. In addition, no dependencies graphs were generated for O-B problems, though some of them are mentioned in the text. The graphs used show two visuals, one for a construction process map and another one for dependencies. Construction process maps reveal the whole BM picture for each identified problem, by displaying the links between its four strata – whereas dependency graphs indicate how problems are disseminated across layers and how they link up with one another.

44 Not showing O-B problems.

ECONOMIC		
PARTNERS *– High number of stakeholders with overlapping roles and responsibilities* *– Discoordination between construction actors*	**ACTIVITIES** *– Project process confused for business model* *– Decentralized decision-making* *– Innovation mismanagement* *– Absence of an entity that could orchestrate time horizons and handle maintenance works*	**RESOURCES** *– Fragmented projects*
CUSTOMER SEGMENTS	**CUSTOMER RELATIONSHIP** *– Top-down managerial approach*	**CHANNELS**
VALUE PROPOSITION *– Failing BM leading to abridged growth prospects at industry level* *– Lack of understanding of smart construction process* *– Big companies' resistance to change* *– Constricted approach to value creation* *– Loss of control over quality of works* *– Mismatch between promised and delivered final product*	**COSTS** *– Faulty estimates leading to recurrent budget overruns* *– Scheduled timelines left in tatters* *– Additional costs incurred to integrate a new technology and upgrade asset*	**REVENUES**
SOCIAL		
LOCAL COMMUNITIES	**GOVERNANCE** *– Banishment of any collaborative urban governance scheme*	**EMPLOYEES**
SOCIAL VALUE *– Outdated CVP* *– Broad concerns apropos the livability of assets*	**SOCIETAL CULTURE** *– Misconception of smart city notion*	**SCALE OF OUTREACH**
END-USER *– End-users not involved in the early stages of construction* *– End-users not being co-creators of value*	**SOCIAL IMPACTS** *– Criticism of final product by end-users*	**SOCIAL BENEFITS**

ENVIRONMENTAL		
SUPPLIES & OUTSOURCING	**PRODUCTION**	**MATERIALS** *– Non-sustainable construction*
FUNCTIONAL VALUE *– Lack of smart city initiatives and urban efforts* *– CSR-related topics occasionally conversed for mere marketing purposes*	**END OF LIFE**	**DISTRIBUTION** *– Large amount of logistics, pieces and other resources tied in the process*
USE PHASE	**ENVIRONMENTAL IMPACTS**	**ENVIRONMENTAL BENEFITS**
TECHNICAL		
BIDDING PROCESS	**SMART CONSTRUCTION** *– Smart city projects as disconnected micro projects*	**LEGAL ASPECTS**
TECHNOLOGY *– Too much weight put on technology in the building of smart cities*	**ADMINISTRATION** *– Overlooked organizational innovation*	**HUMAN CENTRICITY** *– Smart constructions as pure technology demonstrators*
OFFERINGS *– Risk of technological obsolescence upon delivery of project*		

Table 5.3. *Breakdown of problems: a sneak peek (source: created by the author)*

5.3.1.1. *Dependencies*

We use the TLBMC for the mapping of problems. From our viewpoint, problems should not be contemplated in isolation, as, say, one economic problem could have repercussions, direct or indirect, on other noneconomic layers of the model. Indeed, problem dependencies could be dotted between layers and among building blocks. The same problems could even sometimes affect several building blocks within the same layer.

The **construction industry's failing BM**, for instance, falls within the model's economic layer, part of the **value proposition** building block. Indeed, a failing BM could lead to worsening offerings that do not match people's real needs and aspirations. This would over time influence other building blocks within the economic layer, **revenue** for instance. This is viable as inadequate value propositions would eventually lead to reduced sales and amplified stocks of unsold units, hence plummeting construction companies' financial returns. Other corollaries could be spotted at the level of other layers too. In fact, a **failing BM** may signpost the existence of a **weakening governance model**, thus affecting the **social layer** of the model, precisely its **governance** building block. Likewise, there might also be causality between a **failing BM** and a **deteriorating production system**, a system that does not, for instance, encourage sustainable construction practices, which implies that the impact of the first economic problem may prolong reaching the model's **environment layer**, its **production** building block in particular.

This analysis suggests that problems in construction today do not only affect one specific model's component – but on the contrary, their impact may extend to reach other components within. Thus, social, environmental and technical repercussions may perhaps supervene from economic problems – and vice versa. **A constricted approach to value creation** is another problem faced in construction today, whose dependencies are similar to the ones formerly described. The same analysis could be used for the explanation of other problems, mainly: **lack of understanding of smart construction processes** and **big companies' resistance to change** (see Figure 5.2 – as you may notice, in this case, end-users are not part of the process and do not partake in the conception of the project).

If we consider another problem, say **faulty estimates in construction** – for instance, the latter obviously falls within the economic layer of the model, within the **building block entitled costs**. As formerly evoked, faulty estimates are often the result of misinformed or incompetent project managers (architects) who make budget previsions that end up being erroneous. Though this problem is economic, it is also of a technical nature, linking to the technicalities of the project and triggering another problem, which is the **maladministration of large construction projects** in general. With that said, faulty estimates may have an impact on the **technical layer** of the model, mainly its **administration building block**. If we were to extend our analysis further, we could say that significant faulty estimates could even lead, in certain cases, to the annulment of construction projects, thus affecting the **blocks, activities and partners** within the model's economic layer. Construction activities would be permanently stopped and money paid by partners, including investors, would have gone down the drain. This is an extreme scenario though. If we were to consider another, more conservative scenario where the budget, though underestimated, remains adjustable, we could assert that construction companies, seeking to preserve their profit margins, would tend to correct their spending so they fit into the budget's envelope. Often, budget cuts are made to the detriment of the environment and social aspects of projects, which means that potential repercussions on the model's social and environmental layers could be expected too, possibly: **unsustainable construction (production block)** and **dissatisfied staff (employees block)**.

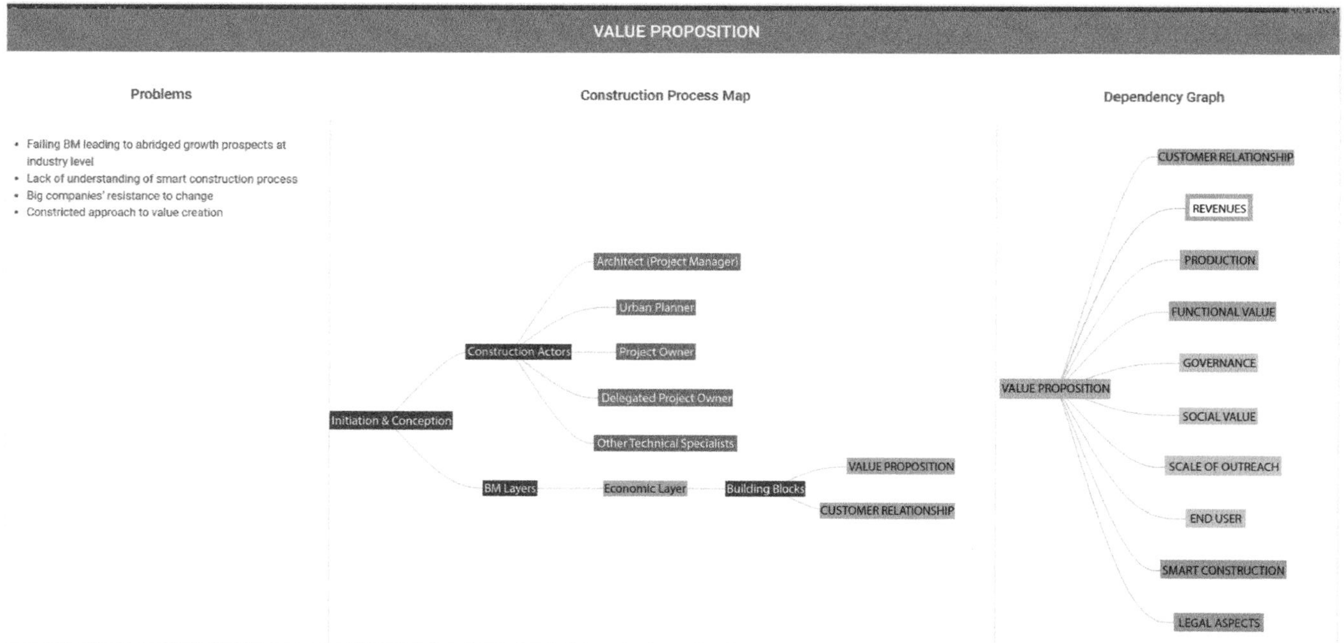

Figure 5.2. *Dependency graph (Failing Business Model). For a color version of this figure, see www.iste.co.uk/karam/general.zip*

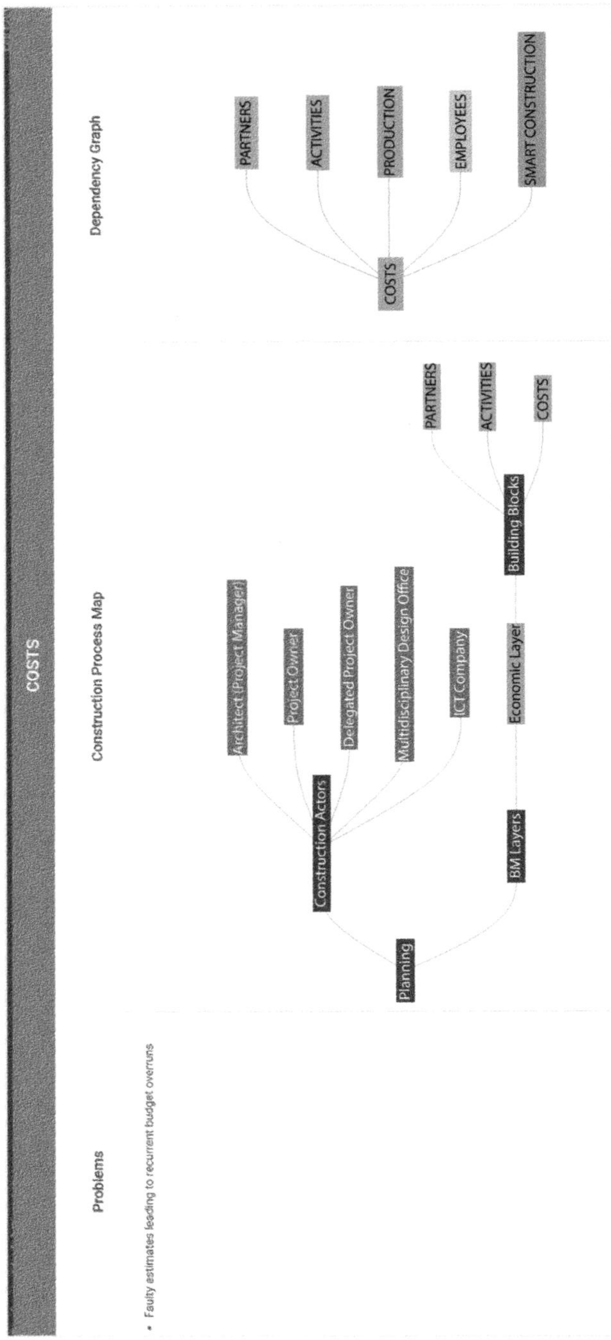

Figure 5.3. *Dependency graph (faulty estimates). For a color version of this figure, see www.iste.co.uk/karam/general.zip*

Our scrutiny could extend even further to cover the little ramifications that may potentially ensue from a given problem; however, the GC BM created for the purpose of this book only highlights some of the most obvious dependencies between layers and building blocks (see Figure 5.3 for a complete overview of both the construction process and problem dependencies).

Moving forward, the **absence of an entity that can orchestrate time horizons and handle maintenance works** is another problem whose repercussions touch three building blocks within the economic layer: activities, value proposition and costs. Often, delayed project deliveries are not exactly happily welcomed by end-users, which can possibly (negatively) influence the company's relationship with its customers. Also, to incur extra unforeseen costs for the restoration of structures will have to be paid by companies (costs block within the model's economic layer). Such snags are often the result of mismanaged construction activities, where the discoordination between stakeholders involved in the process would lead to time overruns and low-quality works that would necessitate huge maintenance works only a short while post-delivery. In the mid to long run, such a problem is likely to put companies' value propositions into question. Consequently, the impact of the said problem extends from the economic to the social layer, affecting its end-user, social value and social impact building blocks (Figure 5.4).

Similarly, the **mismatch between the promised and delivered final product** is another problem that finds its origins in the model's economic layer, at the level of its value proposition building block, yet extends to touch another block within the same layer, customer segments. As detailed in earlier sections, construction actors make promises at the early stages of the construction process but fail to deliver at the end. This is factual as no construction company has control over the entire chain. With the relay passed from one stakeholder to another throughout the construction project lifecycle, the promised product is likely to change stage after stage, especially if no entity is overseeing the process, down to upstream (affecting the model's activities block). Frequently, end-users get a final product that is different from the one expected, leading to their discontent (relating to the customer relationship block). As for construction companies, they time and time again mistakenly presume to know what their customers' needs really are. They deliver walls and roads, whereas end-users are looking for something different. Not all customers are the same, nor do they have the same preferences and tastes. This is a vital piece of information that is often disregarded by construction companies. Hence, what construction companies end up proposing would be lacking the expected value. This suggests that the problem identified has an influence that goes beyond the economic to reach the social, chiefly its end-user, social value and social impacts blocks (Figure 5.5).

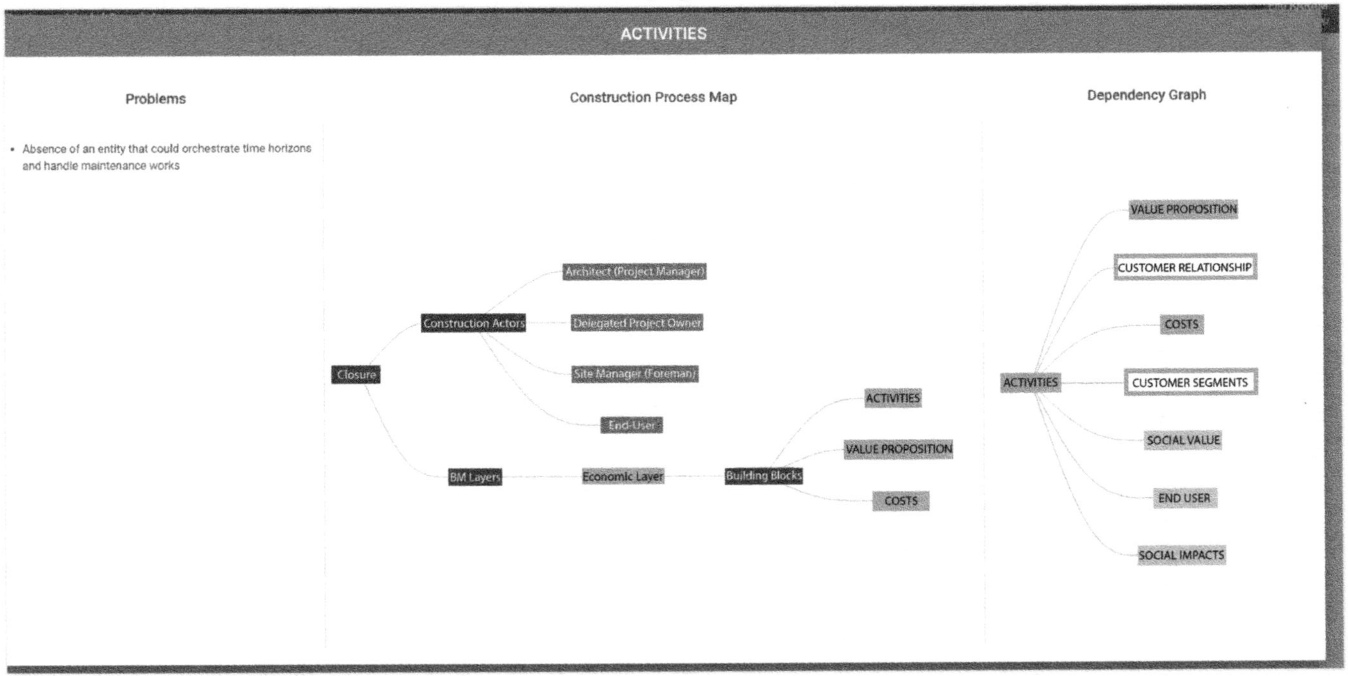

Figure 5.4. *Dependency graph (absence of an entity that could orchestrate time horizons and handle maintenance works). For a color version of this figure, see www.iste.co.uk/karam/general.zip*

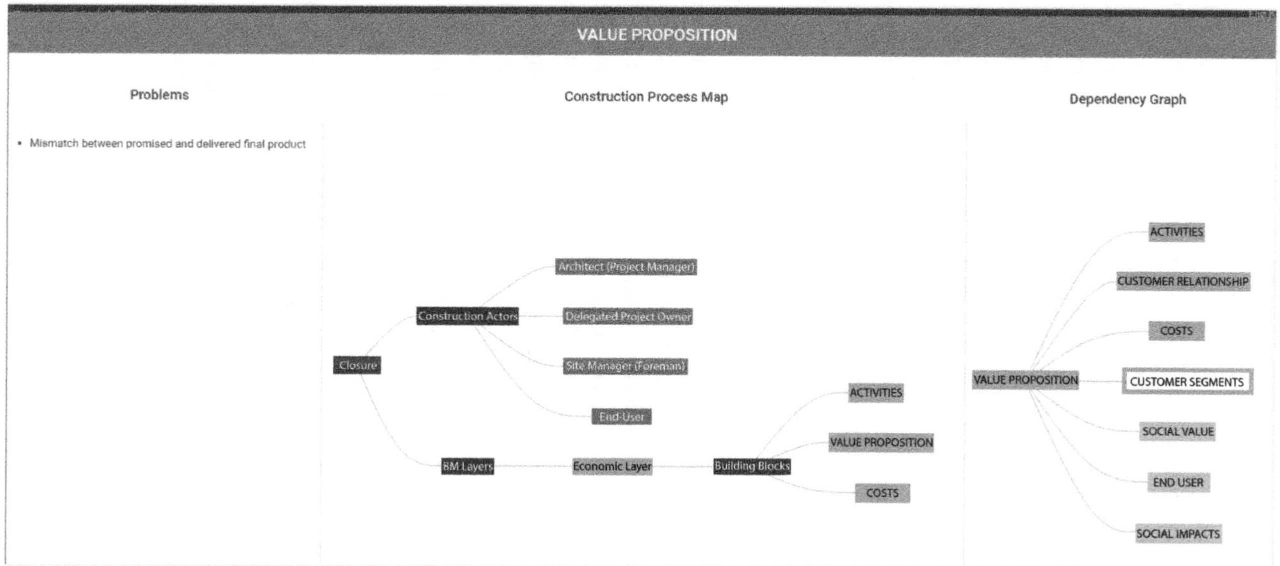

Figure 5.5. *Dependency graph (mismatch between promised and delivered product). For a color version of this figure, see www.iste.co.uk/karam/general.zip*

5.3.1.2. *Additional costs incurred to integrate new technology and upgrade assets*

It is a problem whose impact extends to cover almost all of the model's layers. As its name reveals, it is part of the economic layer – the building block entitled cost. *Ipso facto*, we say, as costs increase, revenues would drop, which means that the revenue building block would also be indirectly affected. To integrate new technology and upgrade assets, we assert, some serious renovation and restoration works would be required[45]; residents would have to evacuate their homes and people living near the project would have to suffer the consequences, loud noises and smashed environs. This implies potential repercussions on both social and environment layers, respectively, their environment impacts and social impacts building blocks. Contemplating things from a different angle, we affirm that other aftermaths could also be marked at the level of those layers, especially since, in such settings, the expected life span of the property would be reduced (use-phase block), and end-users would be frustrated, as they have failed to extract any social value from the property purchased (social value block) (Figure 5.6).

Now let us take an example of an O-B problem that falls within the model's environment layer, its environment impact building block specifically: urbanization. Thus far, we have clearly explained that urbanization has now become a trend. Indeed, with the massive influx of people towards cities, increased pressure has been put on public goods and services offered at city level. All held constant, urbanization could have some serious impacts on the well-being of local communities: poor living conditions, high unemployment and crime rates, and others. Unless big cities turn their territories into smart ones, urbanization would unquestionably have some adverse effects on its inhabitants. Thus, urbanization, which is initially an economic problem, a phenomenon where people flee their homes in metropolises in pursuit of more and better opportunities, is likely to affect the model's social layer, mainly its local-communities building block. Capitalizing on what has been said, the lack of smart initiatives and urbanization efforts is a problem that could have some serious influence on a city's attractiveness and absorption capacity. The latter falls within the model's environment layer, its functional-value building block. However, such smart initiatives are part of companies' value propositions; therefore, this infers that companies' offerings are to a certain extent outdated and distanced from people's requirements, which in turn translate into a lack of social value in the offerings made

---

45 Some useful insights into the matter available at: https://www.khl.com/demolition-and-recycling-international/jacqueline-o-donovan-inside-looking-out-/127864.article (interviews with Jacqueline O' Donovan and other industry experts).

to end-users. With that said, the environment problem in question can possibly influence all of the model's layers, the economic through value propositions, the social through social value and the technical through offerings.

On a separate note, under the old BM, the **amount of logistics and other resources tied in the process of building smart cities** is large. This problem falls within the model's environment layer and directly touches its distribution building block. Nevertheless, it also has an indirect influence on another building block within the same layer: the environmental impact. This is genuine as the higher the amount of logistics and resources held in storage during the process, the higher the environmental impact would be. Looking at things from a slightly different perspective, it would then be clearly noticeable that this problem relates to the model's economic layer as well, in the sense that the logistics of the project could also sway its economics. Specifically, depending on the project location for instance, as well as the channels used for the delivery of resources and materials, distribution costs may significantly increase or decrease. Within this setting, the model's channel building block would be affected by the project logistics. The more we extend our investigations, the more dependencies we will be able to spot. This reveals the dynamism of problems faced in construction today – thus the urgent need to find plausible solutions for their resolution (see Figure 5.7).

For now, it is conceivable that the impacts of problems, even if they fall within a specific layer, could extend to reach – or even instigate other problems in – other layers. This is mainly the case of **outdated CVP**, a problem that falls within the model's social layer, at the level of both its social value and end-user building blocks (Figure 5.8). Outdated CVP relates to inadequate offerings (building block within the technical layer) that do not align with people's real aspirations. A value proposition (building block within the economic layer) does not hold use-value among its layers in the eyes of end-users (relating to the use-phase building block within the environment layer). This problem also sheds light on the fact that construction companies may be using their resources (block within the economic layer), financial and nonfinancial, inefficiently for the production of undesirable artifacts. Other dependencies may also apply; however, they are indirect.

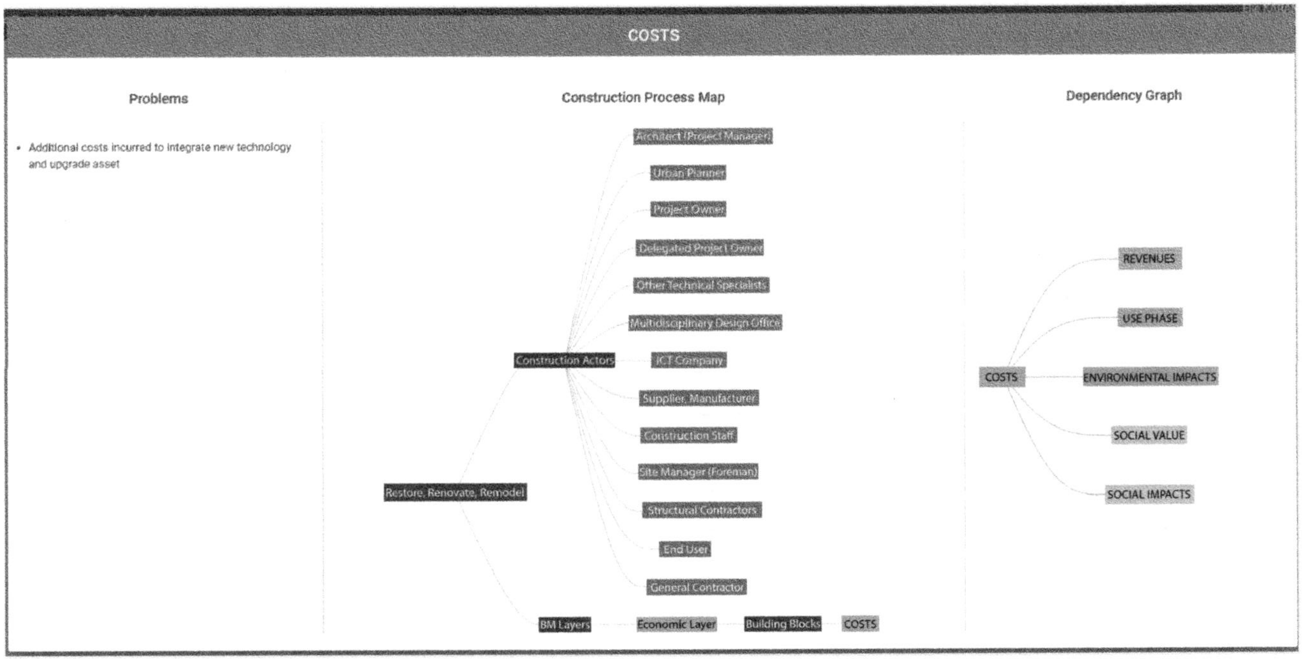

Figure 5.6. *Dependency graph (additional costs incurred to integrate new technology and upgrade asset). For a color version of this figure, see www.iste.co.uk/karam/general.zip*

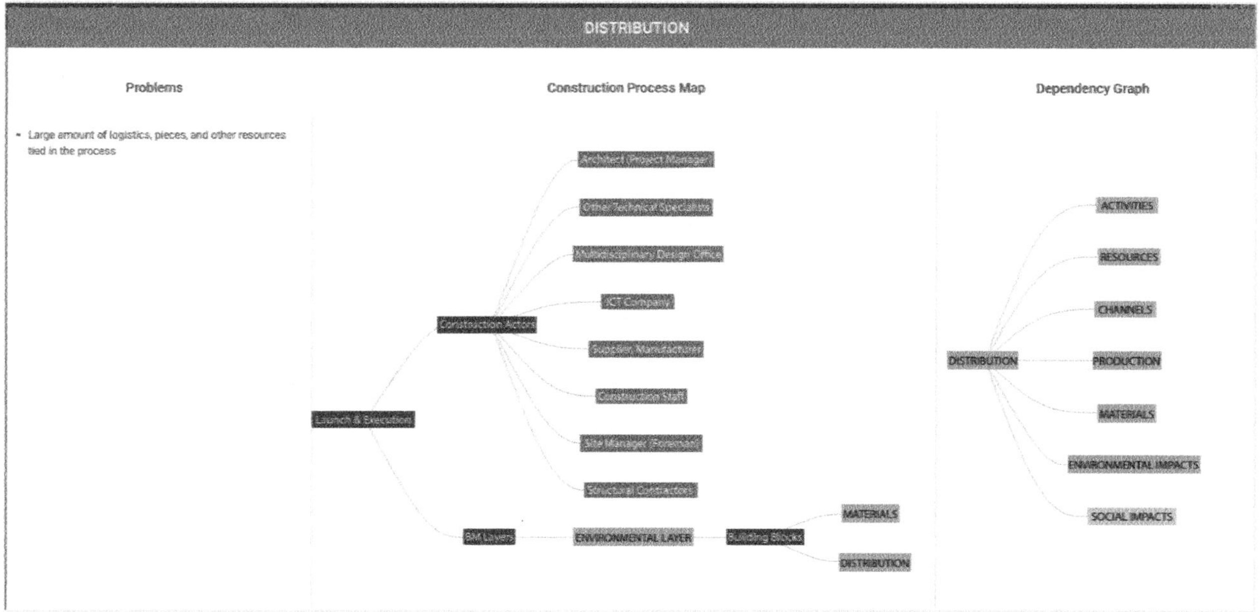

Figure 5.7. *Dependency graph (large amount of logistics, pieces and other resources tied in the process). For a color version of this figure, see www.iste.co.uk/karam/general.zip*

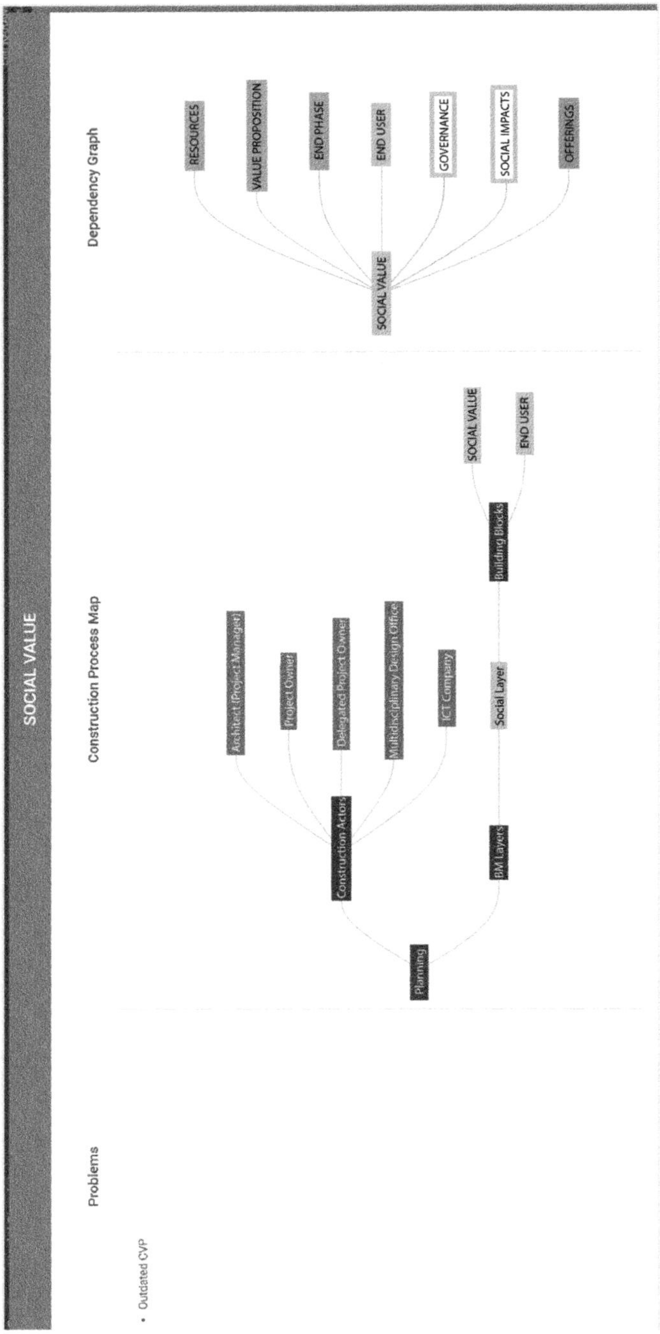

Figure 5.8. *Dependency graph (outdated CVP). For a color version of this figure, see www.iste.co.uk/karam/general.zip*

The **misconception of smart cities problem** relates to the social as much as it does to the economic, environment and technical (Figure 5.9). As reported throughout the present chapter, a smart city is a question of culture and governance – being the two affected building blocks within the social layer. On the economic and environmental fronts, the dependency is straightforward, a smart city is a proposition which would undoubtedly be destined to fail if it does not hold use and functional value amongst its layers. Finally, a smart city is a blend of variety of smart things, among which smart constructions fall within the model's technical layer. We tackle one more social problem, before discussing one technical problem and concluding this section: **end-users not being co-creators of value**. This problem falls within the model's social layer (Figure 5.10), obviously its end-users building block. As discussed, in the realm of smart cities, end-users are not only sources of data but creators of innovation too. Therefore, end-users are resources (or should be regarded as such!) that would help in the building process of smart cities. If they were not involved in the construction process, the offerings made by construction companies would be inadequate and lack value. Indeed, the life expectancy of the final product would be shorter and the utility that end-users would extract from it would be trivial. Ultimately, the said problem would impact the entire customer journey.

Within the same rationality, we assert that, in smart city projects, the **risk of technological obsolescence upon delivery of the final output** is normally high; especially since the life span of technology is shorter than that of a smart city project. Though such technical nuances are important, with the possibility to sway the allure of construction companies' offerings and the economics of their projects, they repeatedly ignore them. This problem falls within the model's technical layer, its offerings building block – and within the economic layer, relating to value proposition. Given the massive gap between the life span of technology and that of a smart city project, whenever technology becomes obsolete, some serious restoration work must be done to integrate new technology and upgrade assets, thus affecting the model's environment layer, its end-of-life building block specifically (Figure 5.11).

Now that the logic behind problem dependencies has been thoroughly explained, we carry on by exhibiting and commenting on the GC BM stage-based construction maps, showing plausible solutions to problems discussed in earlier sections.

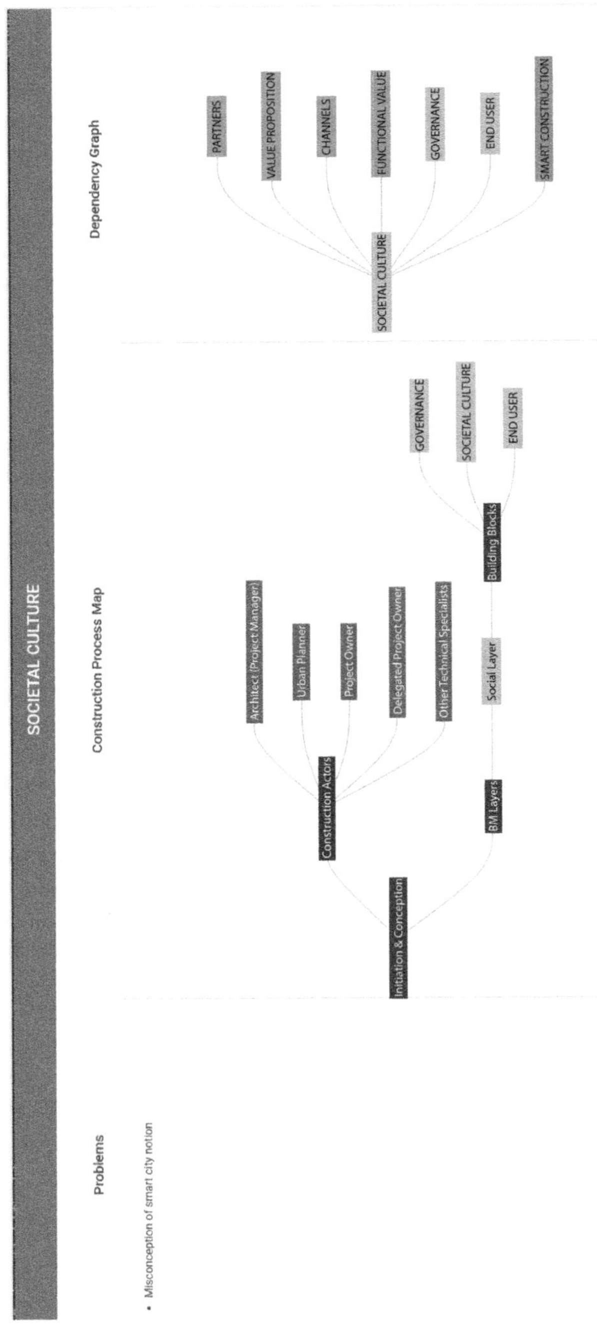

Figure 5.9. *Dependency graph (misconception of smart city notion). For a color version of this figure, see www.iste.co.uk/karam/general.zip*

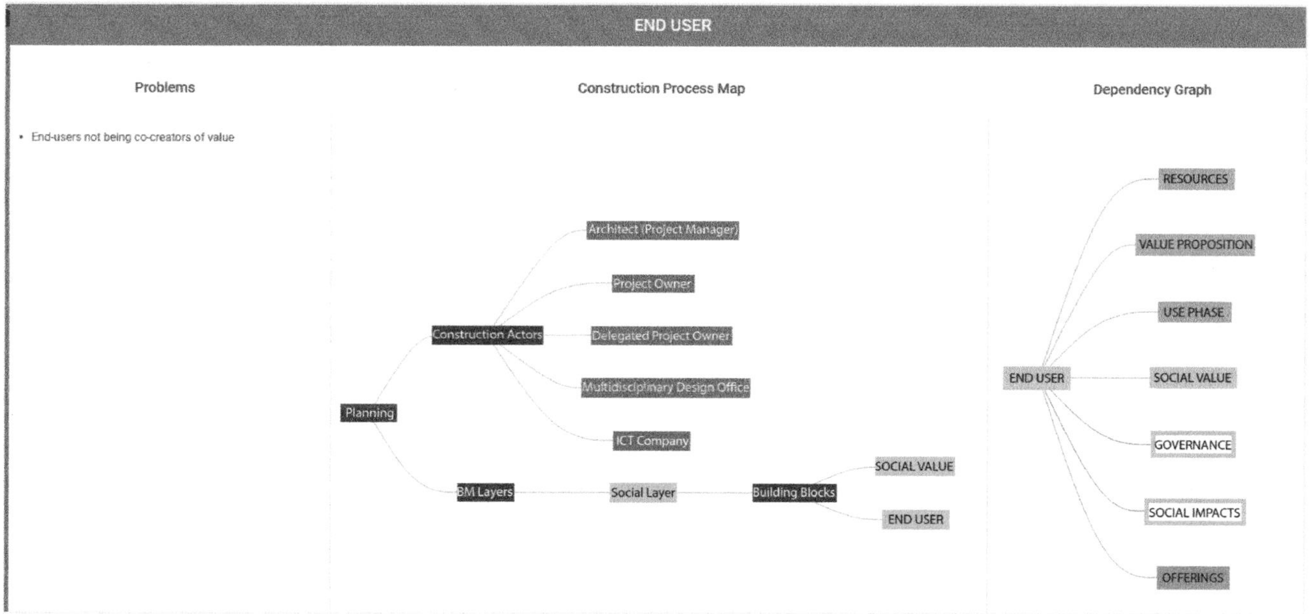

Figure 5.10. *Dependency graph (end-users not being co-creators of value). For a color version of this figure, see www.iste.co.uk/karam/general.zip*

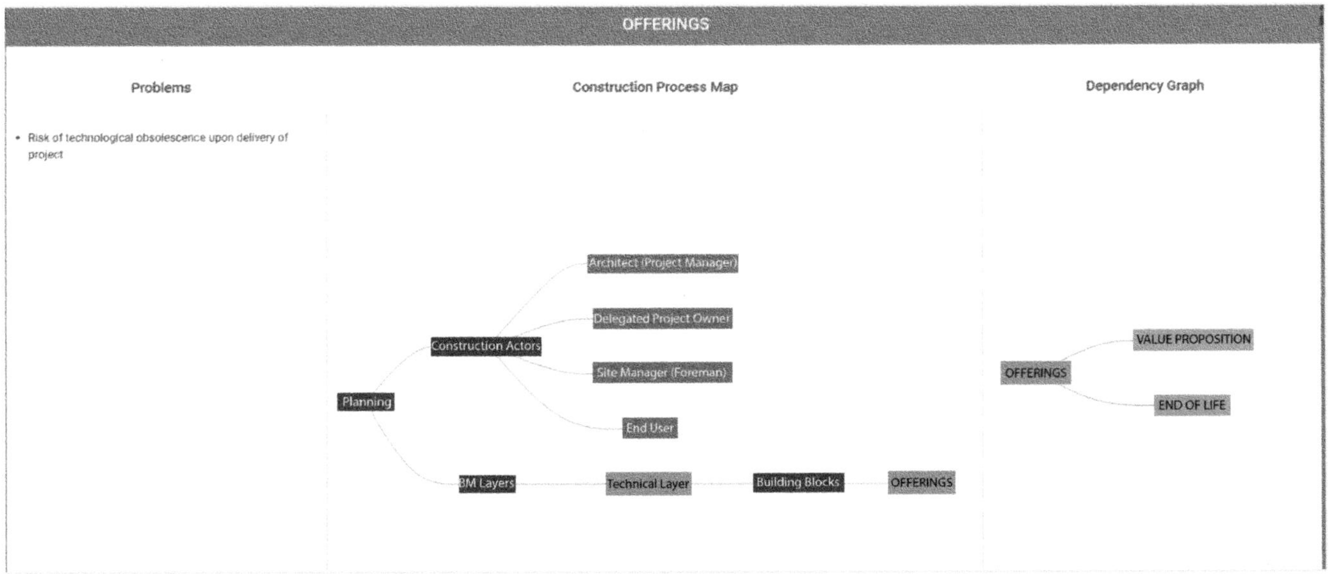

Figure 5.11. *Dependency graph (risk of technological obsolescence upon delivery of project). For a color version of this figure, see www.iste.co.uk/karam/general.zip*

5.3.2. *Breakdown of solutions*

In this section, we spread the solutions suggested over the GC BM layers and building blocks. Now that we are talking about solutions, we use the GC BM in its complete form, accounting for six rather than four strata. Before we deep dive into the narratives and graphical illustrations of construction process maps, we first indicate that solutions were not suggested to match every problem identified, as some problems may not have a solution. In our reasoning, a suggested solution may apply for more than just one problem. For instance, if we were to assign a central operator from the very start of a construction process to manage the entire process and oversee construction works from start to finish, the bulk of problems that usually arise in later stages of the project process, with the possibility of affecting one or more of its dimensions, would then be avoided. This also explains why no dependency graphs were generated for solutions (in contrast to problems).

Moving on, given that the business model simulator that we created for the purpose of this book is capable of generating tenths of graphical illustrations, we proceed by only choosing some of the solutions suggested, and displaying and commenting on their respective construction process maps.

Knowing that our GC BM is a stage-based BM, we run our analysis sequentially by firstly inspecting two solutions suggested for the proper initiation and conception of large construction processes: adjusting offerings to the exact needs and requirements of end-users (within the model's economic layer) and administering smart developments following a bottom-up managerial approach (within its social layer).

Markedly, the construction process maps shown below display inherent links between the BM's various strata, going from stratum 1 to 6. In Figure 5.12, we can see that end-users have become an integral part of the process, co-conceiving and co-creating large constructions along with the GC and other key stakeholders (stratum 3). The initiation and conception stage (stratum 1) is connected to programming in the BIM process (stratum 2). Moreover, the solution suggested would supposedly have an impact on three of the model's building blocks, all falling within its economic layer: activities, value proposition and customer segments. Regarding stratum 5, smart city components, a control variable, a solution, for instance, would account for four of its components, two primary (governance and economy) and two secondary (people and mobility). Indeed, by adjusting their offerings, construction companies would be able to secure a higher return on investments, as they would eventually be able to sell their products.

Moreover, given that end-users partake in the conception process, companies are following a resourceful governance model for the building and management of smart constructions. Following the same reasoning, the conception and building of smart cities following a bottom-up approach (Figure 5.13), encouraging the involvement of end-users throughout the entire process, would have some positive impacts on the project's social value. In addition, through real concessions and hearing sessions for instance, social acceptability for the project would heighten and local communities (referring to potential buyers) would be happy to invest and settle within it. In such a case, the living component of a smart city (among others) would be ensured.

Hence, we elucidate, in construction projects, most operational problems occur in the very early stages of the construction processes and snowball over time to create other problems affecting some or all of the project's dimensions. Thus, by using an adequate governance model from the beginning, many of the problems that often arise during more advanced stages of the process would probably be avoided. Most of the incremental changes suggested by the GC BM concern the very early stages of construction processes.

Moving on, while sticking to the initiation and conception phase, we briefly discuss another solution that falls within the GC BM's economic layer, its customer relationship building block specifically: ensure involvement of end-users in all stages of the construction processes. As shown in Figure 5.14, at the conceptual design level of the BIM process, architects would handle the project's drawings, ICT companies the design of its digital infrastructure, and multidisciplinary offices the approval of project specs. In addition, under the GC BM, each stakeholder would have their own *raison d'être* – and large construction projects would be conceived and built collaboratively rather than individually. Moreover, under the GC BM, we advocate that large construction projects are better off ran, with higher chances of success, by a central entrepreneurial entity. We carry on by stating that the solution suggested in this case would primarily touch upon two smart city components: governance and economy. As previously evoked, by espousing a governance model that encourages the involvement of end-users in construction processes, the project economics would be expected to improve, and so would the project's estimated financial returns.

Skipping to the planning stage, at the detailed-design level of the BIM process, we tackle another suggested solution for the resourceful building of smart cities: allowing for customizability (portfolio diversification).

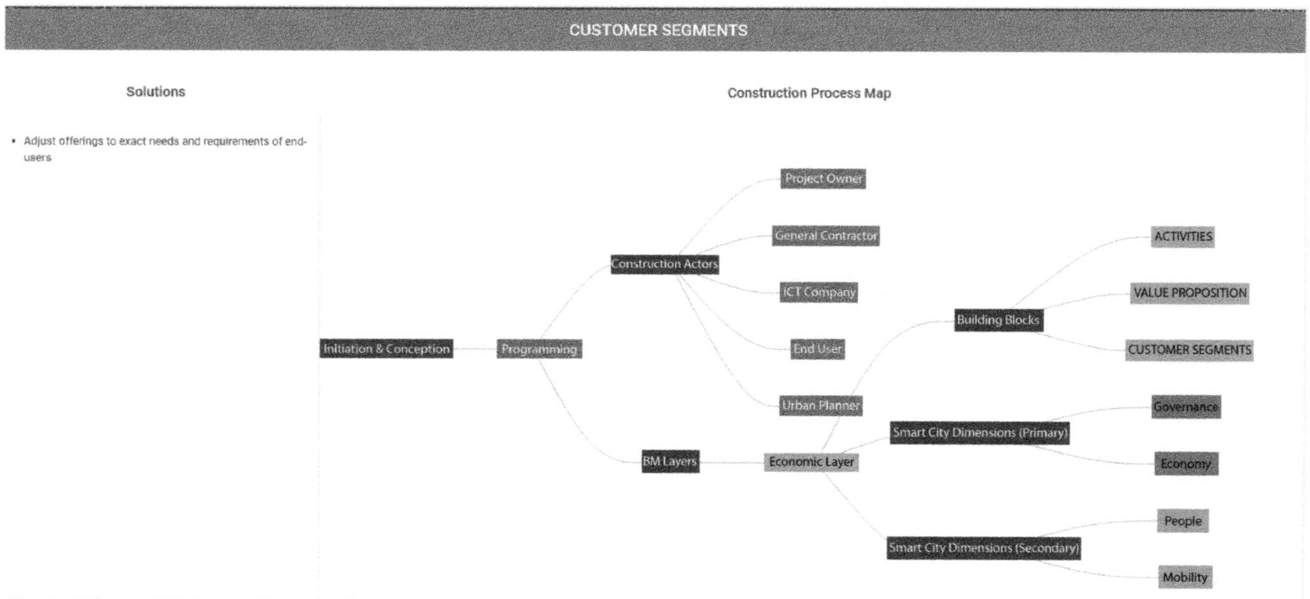

Figure 5.12. *Construction process map (adjust offerings to exact needs and requirements of end-users). For a color version of this figure, see www.iste.co.uk/karam/general.zip*

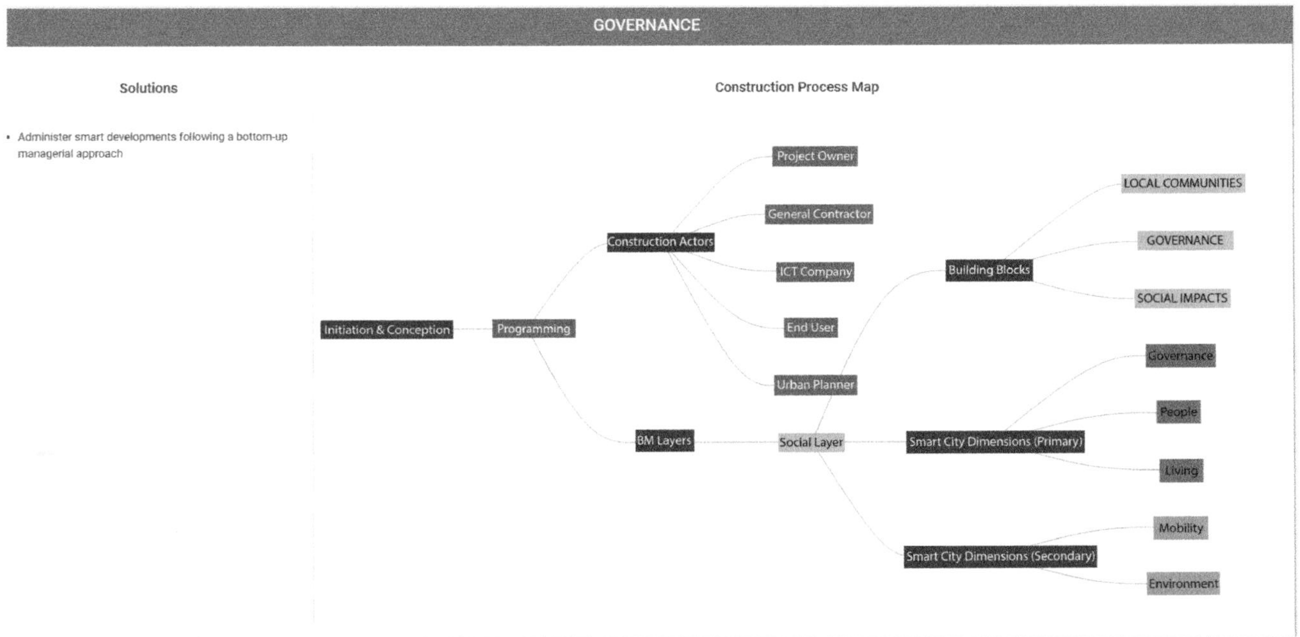

Figure 5.13. *Construction process map (administer smart developments follow a bottom-up managerial approach). For a color version of this figure, see www.iste.co.uk/karam/general.zip*

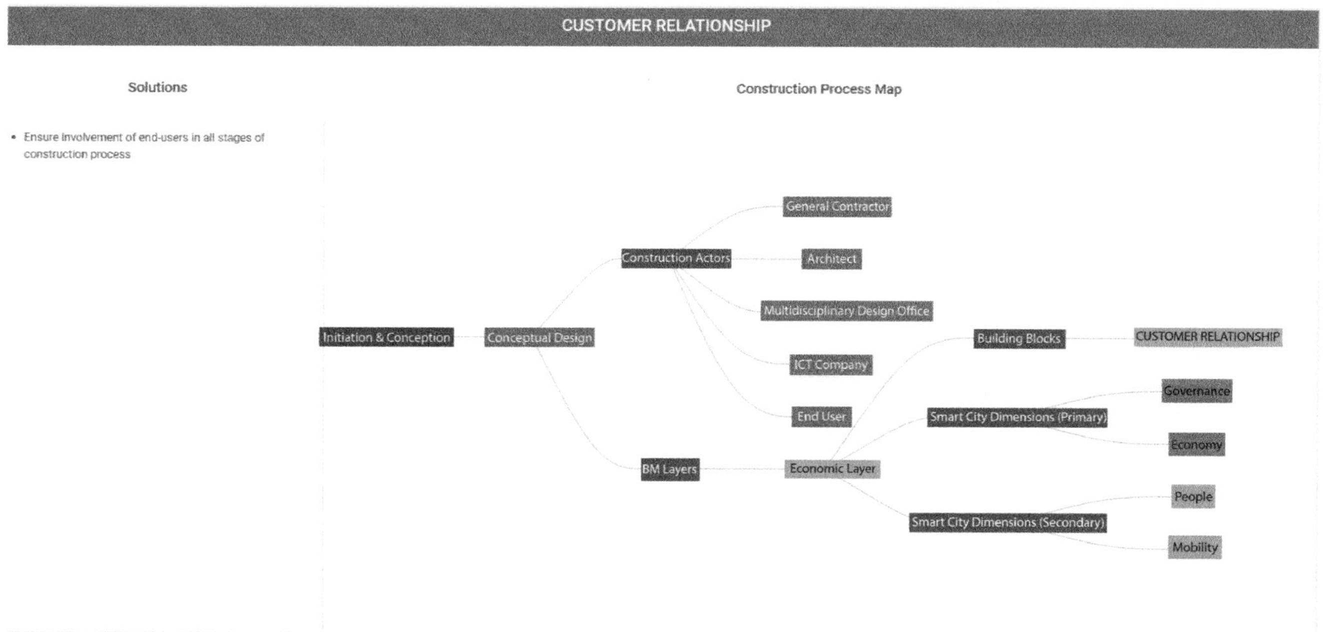

Figure 5.14. *Construction process map (ensure involvement of end-users in all stages of construction process). For a color version of this figure, see www.iste.co.uk/karam/general.zip*

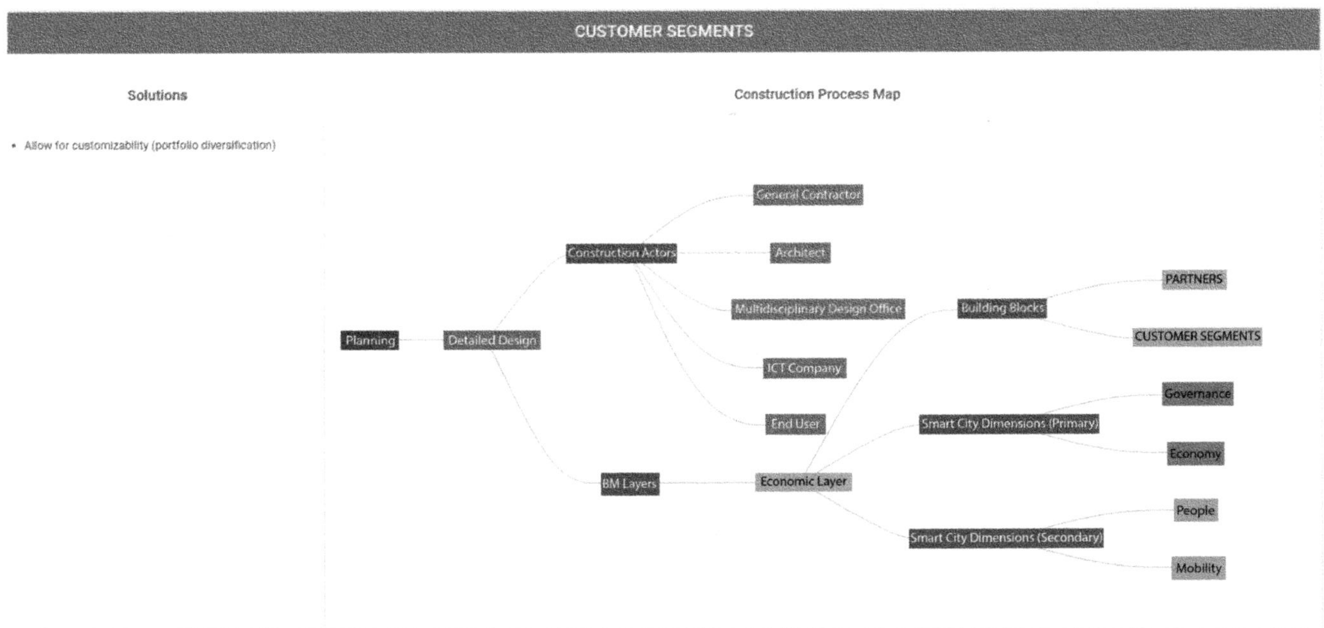

Figure 5.15. *Construction process map (allow for customizability). For a color version of this figure, see www.iste.co.uk/karam/general.zip*

Following the same reasoning as above, the appropriate construction actors intervene at this stage, directed by the GC, in the presence of end-users to settle upon the detailed design of the smart city. This solution falls within the GC BM's economic layer, its partners and customer segments building blocks in particular. This is straightforward, as under our envisioned BM, end-users are active partners who partake in the entire construction process. Besides, construction companies, seeking to enhance the size of their respective market shares, would positively opt for customization. This is a reasonable corporate tactic – as not all customers have the same needs and wants. Thus, through portfolio diversification, construction companies would be better positioned to expand their business activities and serve a larger portion of the market. Looking at our control variable, we could openly avow that the solution in question can possibly have a marked influence on a smart city's governance and economy components (primary) – and, on other ones too, people and mobility (secondary) – to a lesser extent (Figure 5.15).

Now that we have commented on some of the solutions suggested for the resolution of problems faced in construction today, we conclude this section at this level, simply because the same reasoning, as described above, applies to the remainder of solutions (see earlier sections for a detailed discussion of the solutions suggested).

5.4. Conclusion

Throughout this chapter, we plainly identified and discussed the dynamics of problems faced today in construction and suggested reasonable solutions for their resolution. Though the recommended improvements are specific to large or smart construction projects, they could also apply to other projects of relatively smaller sizes.

For the building of smart cities, we opined that the long run rather than short run – collectivism rather than individualism – bottom-up rather than top-down managerial approaches – centralized rather decentralized decision-making processes (among others) should be endorsed, otherwise, any attempt to convert our cities into smarter ones would be destined to fail. Such improvements, we showed, are made possible through the GC BM. The GC, a key role in our model, acts as an integrator who brings all of the builds of the city together, thus creating a coherent system that would enhance both its attractivity and economic power. Keeping in mind that the GC role is relatively new in France; the espousal of our innovation would not happen right away, overnight, as some further real-world tests should be performed to confirm its validity and sustainability. As we see it, the GC BM is a step forward towards building a made-in-France smart city business model.

5.5. Appendix

5.5.1. *Interview technique*

As put by Bogdan & Biklen (2006), an IDI is a qualitative data collection technique that allows the interviewer to center the interviewee's speech on a predefined set of topics that are recorded in a discussion guide.

More specifically, an IDI does not enclose the interviewee's speech, but, on the contrary, it allows him to share his personal views and thoughts vis-à-vis matters addressed in the guide. This type of interviewing, we say, enables interviewees to express themselves freely within the framework of a central topic defined *a priori*. The importance of IDIs lies in the capacity of the interviewer to manage the flow of insights passed on by the interviewee while, at the same time, granting them all the autonomy, time and space needed to respond objectively and openly to all the questions asked.

For our study, the use of the qualitative research method has enabled us to gather descriptive, narrative and macro insights into the topic at hand and to build some concrete knowledge of a practical, hands-on system, explicitly: the building and managing of smart cities and the roles that GCs could play in this regard.

While the recruitment of participants was mostly done via telephone, in some other cases, prospective respondents were approached through written communication (i.e. e-mails).

The interviewees were very receptive and responsive; they showed lots of interest in the study, and enthusiasm and willingness to participate in it; they openly shared their lifelong knowledge vis-à-vis various topics of interest. On average, each interview lasted for about 90 minutes. The IDIs were carried out face-to-face – via PAPI (Pen & Paper Interviewing) – based on prescheduled appointments and using a specialized discussion guide. They were recorded and transcripts were made to use the information.

5.5.2. *Interviewees*

Initials	Job Role Title — (Entity)
B-G	Business Development Manager (a leading French construction company)
C-R	French university professor

D-B	Director General of a social housing rental house
F-C	University Professor, Researcher, Sociologist, and Urbanist
E-N	Development Director in a French *Grande Ecole*
P-P	Ex -General Director (a leading French construction company)
D-T	Economist in a big urban planning cabinet
S-P *et al.*	Chief Operating Officer (a public developer)
P-B	Strategic Consultant (independent) A university instructor and researcher
O-A	Development Director (a leading French construction company)
AF-H & B-S	Innovation Director (a leading French multidisciplinary design office)
J-N	Operations Director (a French multidisciplinary design office)
A-T *et al.*	Chief Operating Officer (a French multidisciplinary design office)
V-B	Executive co-director, Master's program in social housing at a French university Secretary General of 2 associations related to social housing

6

Discussions

Given the title of this book, it was necessary to tackle three distinct subject areas, namely: smart cities, general contractors and business models. In the following sections, we clarify our contributions to the available academic literature that relates to these themes.

6.1. Theoretical and methodological contributions

We first list our contributions to the topic of smart cities, followed by that of general contractors, and finally of business models.

6.1.1. *On smart cities*

6.1.1.1. *Definition of smart cities*

Today, definitions of smart cities are numerous and variable (Cohen 2012; Kitchin 2014; De Jong *et al.* 2015; Simard 2015; Eremia *et al.* 2017), are often disparate and can lack meaning. For some, smart cities are places that can guarantee higher quality of life. For others, however, smart cities are eco-friendly places, focusing on the environmental aspects of a city; monitoring and reducing pollution and GHG emissions, as well as energy and water consumption. Others regard social equity, together with a greener living environment, as key considerations in the development of smart cities. In recent developments, new ideas have emerged, with the smart city concept being defined as a combination of people, activities and environment. Places in which a dynamic balance between economic, environmental and socio-cultural development objectives could be achieved; outlined by local governance systems characterized by increased citizen participation.

Today, "smart city" has become a catchword, drawing increasing attention among research agencies, universities, governments, officials and specialized companies. All over the world, smart city programs have been promoted and have therefore proliferated; market potential for smart city solutions is set to be substantial. As we have mentioned on a number of occasions, smart is a word that is used in construction to enable companies to sell their products faster at higher prices; this often creates misconceptions about smart cities among end-users, and sometimes even among industry experts. Today, although the concept of the smart city maintains a favorable position, implementing initiatives in countries with vastly different needs makes it difficult to establish shared definitions and common trends at the global level. That is to say, there is no clear-cut definition of the term smart city, nor any general agreement on what its inherent features are. In Latin America, for example, smart city projects tend to be weighted towards the improvement of security, local government management and mobility, whereas in Asia, the emphasis is instead on the improvement of infrastructure and mobility; there again, in Europe, the focus is on boosting efficiency in terms of the public offering, establishing a socially inclusive culture and improving the well-being of citizens.

Our contribution here is in the form of a comprehensive definition of smart cities: "A smart city is a city that is conceived, developed and run by the government and the people, both smart, for the sake of promoting citizens' well-being and ensuring a better quality of life for all". Clearly, our definition omits the GC as we believe they would act as an out-of-sight mediator, standing between public developers who own large construction projects and people who would eventually invest their money to buy properties in those projects. Frankly, it is neither the government nor the people who would build smart cities but construction companies. However, for this to be conceivable, a centralized decision-making process should be espoused, with communication channels remaining open at all times between all stakeholders at the outer edges of the construction value chain. In our views by defending the interests of all stakeholders involved in construction projects and creating bridges between them, maven GCs, as central operators, would be able to ensure the successful execution and delivery of smart projects in general.

6.1.1.2. *Smart city components*

The smartness of a city is indicated by the set of physical and legislative infrastructures that support economic development, ensure social inclusion and allow environment protection. The smart city model (in particular, that of Giffinger *et al.* (2007)) is said to be the most illustrative in terms of its capacity to demonstrate and exhibit the strategic components of a smart city. Nevertheless, since these components were established around 10 years ago, numerous authors have referred to them year after year to develop more elaborate frameworks and strategies, namely Cohen's (2012) Smart City Wheel. For us, a synthesis of the works of Giffinger *et al.* (2007), Cohen (2012) and Simard (2015) allowed for a better understanding of the relevance of smart city components. Indeed, in our study, we stated that, although all smart city components are important, some of them are more important than others; this is why we chose to split them into primary and secondary components. Depending on the type and scale of the construction project, some components seem to have more weight in comparison to others. Overall, we believe governance and people are the key components of smart cities.

Furthermore, we believe, our contribution to the literature on smart cities went beyond the theoretical to reach the practical. In fact, we made use of some real-life smart city examples to elucidate our viewpoint. Songdo (South Korea) was put forward as an example of a failing smart city model and Copenhagen (Denmark) was given as an effective one. While the former focused on technology for the building of the city, the latter instead concentrated on human-related aspects. Within this framework, we continued by affirming that a smart city is a number of smart things: governance, energy, infrastructure, transportation, healthcare, building, etc. Looking at the world's experience in this domain, we could clearly see that some components are now being prioritized over others.

Currently, the components that are drawing most attention globally are governance, energy and infrastructure. Nevertheless, our research findings show that "*people*" is the key factor to the successful building of smart cities, and that it is only by ensuring parity between the various dimensions of construction projects that those projects would end up being smart. We stated that this is only possible through smart governance, the assignment of a central operator who, in collaboration with other stakeholders (including end-users), would oversee and manage smart construction projects effectively. Finally, we have showed that the building of a smart city is a dynamic and evolving process in the sense that what is thought to be central today could lose ground over time. Year after year, new components come in and others go out. More precisely, though the environment or mobility, for instance, is not currently a primary component of smart cities, we expect it to be in the near future. In addition, while mobility, for instance, may possibly rank first in Copenhagen, this would not necessarily be the case in London, Amsterdam or even Paris.

In our GC BM for smart cities, the "*smart city component*" was used as a control variable to ensure suggested solutions took account of the requirements of smart cities. For the dissemination of problems identified in the construction industry today, the latter was not part of the four-stratum BM used to this end. Alternatively, problems were spread across the layers of the TLBMC (economic, social, environmental and technical), which actually enabled us to grasp how and why construction activities today do not align with smart city requirements. As mentioned, our choice of the TLBMC was influenced by the inherent similarities between its layers and the components of smart cities, as illustrated by Cohen (2012). For the most part, problems identified related to two components of smart cities: governance and people. This is why narratives such as "with smart cities, it is a question of urban governance that should be developed using a human-centric approach" can be found throughout the entire text. Even by looking into the global smart city market and how it is segmented across various categories of smart things, we can clearly see that smart cities are primarily a question of governance and people. Whenever these two (primary) components are safeguarded, the remaining (secondary) components will automatically follow. Finally, we recommended that in France, the focus should now be on working out public–private partnerships to encourage smart city initiatives at national level, as well as encouraging the involvement of end-users in large construction processes, from initiation to closure.

6.1.1.3. *Are all cities destined to be smart?*

The smart city concept is subjective. Indeed, we stated that there are no right or wrong smart city models, as the meaning of a smart city is likely to change from one place to another, across cities and countries, based on their cultures and traditions. For this reason, we acknowledged that by using their own resources and abilities effectively, refining communication infrastructure and web-based services, allotting ample financial resources and adopting appropriate funding schemes, and creating apposite value chains that ensure fair interplay between stakeholders, all cities around the world could become smart; at different rates perhaps, and in their own form.

6.1.2. *On general contractors*

6.1.2.1. *General contractors, central operators in the new construction value chain*

There is a growing shift today from an old to a new value chain in construction where a central operator would be entrusted to handle the management and running of large projects, from start to finish. Nonetheless, in the literature, there is no mention of who could play this particular role. To our knowledge, today, in France at least, (incompetent) public developers or assigned architects are fulfilling this role. Yet, things do not seem to be moving in the right direction, as the number of problems faced in construction seems to be on the rise. For this reason, we thought of endorsing a GC for this vacant position. Using a basic analogy, a central operator is the equivalent of a *strategi* under Copenhagen's smart city model.

By definition, a GC is a stakeholder – an entity or a person – who supervises daily activities at the construction sites and ensures all construction works are carried out properly. Moreover, a GC supplies the labor, building materials, machinery, raw materials and all other kit necessary for the successful completion of construction works in general. To us, GCs are mavens in construction and are therefore qualified to handle the management of smart developments. As its name suggests, a central operator is a machinist who, if given the requisite power and authority, has the ability to competently oversee construction works and manage teams involved in the process. Yet, "*central*" refers to the centralization of decision-making, meaning the GC would have the final say in project-related matters. In English-speaking economies, the GC is frequently regarded as a central operator who provides tailored solutions to clients and expertly manages worksites, tradespeople, deadlines and budgets. Moreover, we explained, given GCs do not always have the expertise required for completing all construction works by themselves, they tend to maintain a network of specialist subcontractors who can help them complete the work. Furthermore, the GC role is interesting in the sense that the person who assumes it has direct and regular contact with clients to keep them informed about progress, as well as with other construction actors throughout the whole construction project lifecycle. With that said, we believe GCs to have the perfect profile to fill this vacant central operator role, especially in large construction projects.

6.1.2.2. *Rogue analogies*

A rogue analogy was made to better exemplify the role of GCs (as central operators) within the new construction value chain. GCs were compared to music conductors; we also illustrated the rather astonishing similarities between the two. We also asked the following questions: Why are orchestras structured the way they are? Why do the oboes and tubas not sit at the front? Why do flutes and violins not swap positions? Why do trombones and French horns not sit right up front with the music conductor? In fact, we affirmed that there is a good reason for why pretty much everything we experience in life is the way it is, and the orchestra's seating arrangements are no exception. To better elucidate our idea, we referred to big orchestral arrangements (denoting large construction projects). Until the 20th century, the violins, first and second, were seated opposite each other. However, a few years later, the conductor Leopold Stokowski tried something new, seating the orchestra in every conceivable way in an effort to find the perfect amalgam of resonances. He eventually came up with a new, better seating arrangement. At first, his suggestion did not stick but a couple of years later, the so-called Stokowski Shift gained serious momentum and became popular around the world. In our view this is in fact the responsibility of the central operator who, like music conductors, tries to bring all construction actors together in order to create beautiful melodies rather than mishmashes of conflicting sounds.

The GC role (like the Stokowski Shift!) in France is new or nonexistent and current market players, we assume, would not welcome such a proposition despite it being reasonable. Thus, although our suggested BMI may not be adopted today, we hope that it will be in the near future. In our view, the GC BM is innovative in that it identifies problems that are faced in construction today and suggests pertinent solutions for their resolve. Problems are identified per construction stage and split into four different categories. Hence, through the GC BM, which enables problems in construction to be broken down, companies would be able to circumvent all types of snags that will likely hinder their business performance in the medium- to long-term. Within this context – and under the GC BM for the building of smart cities – our contribution lies in the endorsement of the GC role, which is practically nonexistent in France. GCs are popular in English-speaking economies but not yet in Europe. As explained earlier, we believe the GC is perfectly equipped to fill the central operator role in the new construction value chain suggested by Vrijhoef and Koskela (2000). Being a maven in construction, he has sufficient knowledge and know-how to handle and manage large construction projects. Moreover, we stated that GCs, along with end-users, are at the heart of solutions to the problems faced in construction today. Again, we emphasize that smart city projects are best managed by one entity, a maven decision-maker or a *strategi*, based on a centralized governance scheme. Copenhagen's experience in this domain is proof that such a governance model is likely to succeed if it is well implemented. However, for this to be viable, GCs must first have the ability to play this role. Smart cities are connected chains and any flaw in or damage to any part of the chain would undoubtedly affect the entire chain. Hence, GCs would play the role of connectors, bringing together all stakeholders involved in the process and thus, linking the entire construction value chain, down to upstream.

What is more, the "*ensemblier*" concept, which already exists in France, is not the right solution for the various problems encountered in construction today. Let us take the case of Bouygues for instance, a big market operator with lots of subsidiaries whose activities cover almost the entire construction value chain. Bouygues is now facing many challenges handling and managing the work of its own subsidiaries as all of them have the ability to make key decisions that often cause snags and hitches. Why? Simply because, within a project, if more than one decision-maker exists, the odds of operational problems occurring are high. This reflects the absence of centralized decision-making which is in line with Flyvbjerg's (2014) findings: in the case of large and lengthy construction projects, decision-making should be put in the hands of one entity and the management of smart developments preserved throughout the entire construction project lifecycle.

With that said, the GC BM we are proposing constitutes a plausible solution to such problems. Indeed, in our time, big operators like Bouygues are very well equipped to advocate for this new GC role in the running of smart developments. Furthermore, for GCs to fulfill their roles effectively they first must be given the required authority and power to do so. Accordingly, they should act as co-owners of construction projects rather than mere stakeholders.

6.1.3. *On business models*

6.1.3.1. *Definition of a business model*

Despite the recent deluge in the literature on BMs, we believe there is still no clear definition existing today (Chesbrough and Rosenbloom 2002; Afuah 2004; Casadesus-Masanell and Zhu 2013; Massa and Tucci 2014; Baden-Fuller *et al.* 2017; Foss and Saebi 2017). While for some authors, the BM constitutes a key vehicle for innovation at the corporate level, others see it as a representation of how a company generates, conveys and captures value. From our side, we have contributed a further definition to the existing literature:

> A BM is an architecture of the product, service and information flows, including a description of the various business actors and their roles; a description of the potential benefits for the various business actors; a description of the sources of revenues.

6.1.3.2. *A business model for the building of smart cities*

We have demonstrated that BMs could be described as *stractics*, an assortment of corporate strategies and tactics. The rationale for our aforementioned definition of BM is quite straightforward; companies typically start by choosing a logic of (strategy for) value creation and value capture, which would be reflected in their BM. They then make tactical choices that are driven by their intent to maximize stakeholder value. Therefore, sequentially, strategy comes first and tactics follow, and the BM falls in between the two. With that said, we simply used the expression *stractics* to describe BMs because they simultaneously reflect company strategy and influence company tactics. Within this setting, we have also emphasized, albeit indirectly, that generic BMs do not fit the construction industry very well. If BMs are to be useful we believe they must be carefully divided into strata to allow construction companies to keep track of the works at every stage of the process. Therefore, the practicality of generic BMs in the construction industry comes into question. In terms of our GC BM, it is multifaceted in the sense that it is both solution-based and multisided. It is also dual-use because it consists of a BM within a piloting tool. Additionally, it shows which construction actor intervenes and at what time. In the world of smart cities, it allows for problem-solving in construction and the suggestion of plausible solutions for their resolution. The GC BM is a corporate BM that construction companies may use, along with other technologies available in the market, to better track and manage their construction projects. Using six different strata could help construction companies track problems as they arise, at each stage of construction. It shows the dynamics of problems in construction according to their type – economic, social, environmental or technical – and shows links between them. As it is operational, the GC BM could be used for simulative purposes, too. Using the business model simulator that we have created with the help of a developer using specialized IT software (Angular 2+), dependency graphs (between problems) and construction maps (for both problems and solutions) could be generated. The GC BM is therefore innovative in that it allows links to be made between the various layers and building blocks of the TLBMC, something that has still not been sufficiently addressed to date.

Conclusion

Throughout this book, we have shown that connectivity is key – and that functional geography has taken over. This new reality has reshaped and wired connections between countries. It has also helped to ease political tensions, mitigate environmental risks and uphold economic growth. Additionally, we have indicated that people tend to migrate to big cities in search for more connectivity and opportunities. Indeed, urbanization has become a trend, transforming big cities into megacities – and megacities into smart cities. Long story short – "*smart cities*" are a combination of smart governments and smart citizens, tied together using smart technologies. Literally, today, big cities are being revamped to resemble smart villages. For the construction of smart cities, we asserted that there would be a need for more adaptability and flexibility. Through the implementation of a new business model, a fair interplay among construction actors could be assured. Their works would be synchronized and their interventions perfectly cued. However, for this to be achievable, a central operator is required. Raising Copenhagen's case at this point, this operator, a person or an entity, is supposed to be a *strategi* who could expertly orchestrate complex smart city projects.

Moving on with our investigations, we have shown how the construction value chain has evolved over time, with the upsurge of innovation in construction paving the way for the intercession of a new actor – the so-called *central operator*. The latter would act as a music conductor, harmonizing the performance of all industry players so that large-scale, smart developments end up being impeccably built. They would synchronize construction works and decide who does what and when. We have also established that it is unimaginable that a single value chain can fit all types of projects, hence the need to conceive new ones. Equally, we have evidenced that business models, being more perpetual than construction project processes, are grandly decisive of companies' long-term business success. Also, we have ascertained that GCs are both mavens in the field of construction – and connectors, possessing their own networks of backers. Pretty much like music conductors, GCs are frontrunners, with the ability to play various roles and hold different positions athwart the construction industry's value chain. All other things held constant, GCs are best equipped to play the role of central operators. Hence, it is within this setting that we thought of designing a new business model for the building of smart cities, a BM where decision-making is centralized in the hands of a maven GC.

Our suggested GC BM is solution-based and human-centric. Human-centric because the end-user is at the heart of it and solution-based because it suggests plausible solutions for operational problems encountered in the construction today. It is also multisided because the GC intermediates between two stakeholders: project owners, located downstream the chain – and end-users, upstream. As for the changes, our GC BM is likely to bring to the table those that are likely to be incremental rather than radical, and of a managerial rather than technological nature. *Per se*, the GC BM built was the result of extensive research carried out on three subject themes: smart cities, general contractors and business models – also, of an extensive problem-solution analysis carried out to assess the current state of the French construction industry's value chain – mainly the ability of construction companies to successfully build and manage smart city projects. Though the GC BM was tailor-made to the exact specificities of the French construction industry, it is adaptable so that other industries and companies elsewhere in the world may also benefit from it.

Little by little, we succeeded in putting all of the puzzle pieces together. The GC BM was created through trial-and-error. First, we had a raw idea that materialized over the years and led to something tangible. The GC BM is labeled as a six-stratum business model as it accounts for six builds. It is a BM within a piloting tool. Through a series of research methodologies, we were able to identify problems faced in the construction today and suggest solutions for their resolution. For the mapping of those, we used the well-known strategic business-modeling tool: the TLBMC – to which we added an extra (technical) layer. Eventually, the data gathered was disseminated across the various dimensions of a construction project. This was made possible using the business model simulator scripted in Angular 2+. In our analysis, we have reported that most operational problems arise at the very early stages of construction processes and snowball over time to create supplementary problems. Indeed, a technical problem occurring in the early stages of the process, if left unsolved, could materialize year-over-year to create a pool of other problems of different types (social, economic and environmental) in advanced stages of the process. Hence, to address and solve those problems, the focus should be on the downstream rather than the upstream of the value chain, the stages where construction projects are programmed – and designed. The dynamics of the problems in construction were plainly discussed throughout our research, and so were the solutions suggested for their resolution.

Although the recommended improvements are specific to large or smart construction projects, they could also apply to other projects of relatively smaller sizes. For the building of smart cities, we opined that the long run rather than short run – collectivism rather than individualism – bottom-up rather than top-down managerial approaches – centralized rather decentralized decision-making processes should be endorsed, otherwise, all attempts to convert our cities into smarter ones would be destined to fail. Such improvements, we showed, are made possible through the GC BM. The GC, a key role in the model, acts as an integrator who brings all of the builds of the city together, thus creating a coherent system that enhances both its allure and influence. Keeping in mind that the GC role is relatively new in France; the espousal of the GC BM by construction companies, we presume, would not happen right away, overnight. Truthfully, as we see it, our suggested BMI is nothing but a step forward towards building a made-in-France smart city business model – one step in a thousand-step journey.

C.1. Research limitations

Though reported by interviewees to be innovative and original, our suggested GC BM remains incomplete in so many ways. First, time and budget constraints. Of course, we would have been able to do a better job if we had all the time in the world to complete the study – yet this is never the case. As with anything in life, there are time boundaries that we should abide by, while endeavoring to deliver the best output possible. Moving forward, among other limitations, we cite those that are research-related – linking to the TLBMC, the business-modeling tool that we used for the successful achievement of our book. Precisely, the large number of building blocks associated with each of its layers. The interconnectedness of buildings blocks, we attest, made the task of dispersing problems and solutions across them complex, which led to some needless redundancies in the analysis. A better way would have been to take the TLBMC and try to shrink the amount of inherent building blocks so that each layer contains no more than five blocks. Moreover, the size equality of building blocks among layers is a bit doubtful. If we had nine building blocks within the economic layer, is it compulsory to have nine within the other layers too? Hence, we admit, a reduced version of the TLBMC would have permitted a more straightforward spreading of problems and solutions (in statistics, data would make no more or less sense if distributed across too many categories, thus the need to find the perfect number of building blocks within each layer – how many is too many building blocks?). On a side note, on the analysis front, a comprehensive economic examination of smart city components, we proclaim, would have brought extra benefit to our research. Also, by, for instance, marking the economic losses in monetary value that could be expected from every problem faced in construction today, our research may have had a higher impact.

In conclusion, a problem-solution matching exercise could have been done – and consequently, dependency graphs generated for solutions – in the same way as we did for problems. This may have given added meaning to our research. By doing so, we would have made construction companies' lives easier, in the sense that every time they encounter a problem, they could straightforwardly look for the adequate solution (one or more). In our case, the GC BM in its present form, they would have to search for it – and choose from several options.

C.2. Research avenues and next steps

Among the research avenues that we would aspire to embark upon in the near future, we thought of the following: come up with clear-cut definitions (one or more) of smart cities in France – definitions that account for the specificities of territories within cities. By finding a common definition, some sort of benchmarking between smart cities around the world could then be performed. Conducting market research on the subject of smart cities to assess customers' needs, decipher misconceptions and identify existing market gaps seems like a valuable thing nowadays, as people have positioned themselves at the heart of smart cities and so, it would be very useful to better familiarize ourselves, as construction companies, with their needs and requirements.

Another interesting idea is to carry out a detailed investigation of the roles and responsibilities of GCs in the suggested BM, as well as in the French construction industry overall. This, we believe, would grant our suggested BM further exposure and validity in the marketplace. Other research avenues include investigating the relationship (if any) between social and smart housing. This is important as a significant share of the construction market in France is reserved for social housing. Should any connection exist between the two, this would open additional doors for the commercialization of the smart city notion. Finally, we also believe that it would be interesting to find some common ground between the private and public sectors so that they can collaborate on the building of France-based smart cities. As recurrently evoked, smart cities are nothing but the result of collaborative thinking and intelligence.

Lastly, concerning the GC BM, the next steps I would take include using the GC BM in the real world to check and appraise its real use-value. Also, its validity and sustainability should be tested. Indeed, we avow, the GC BM must go through a practical trial phase – and its performance must be improved along the way. Once proven valid and sustainable, some marketing work should be performed to commercialize it and encourage its adoption by construction companies. I am also thinking of approaching BIM to present my GC BM to them, to see whether it is possible to integrate it into their technology – thus becoming part of it.

References

3DRS (2015). WinSun China builds world's first 3D printed villa and tallest 3D printed apartment building [Online]. Available at: http://www.3ders.org/articles/20150118-winsun-uilds-world-first-3d-printed-villa-and-tallest-3d-printed-building-in-china.html.

Abdelkafi, N., Makhotin, S., Posselt, T. (2013). Business model innovations for electric mobility: What can be learned from existing business model patterns? *International Journal of Innovation Management*, 17, 1–42.

Abraham, S. (2013). Will business model innovation replace strategic analysis? *Strategy & Leadership*, 41(2), 31–38.

Abu-Jbara, N. (2019). The innovation of a hybrid business model for social enterprises. *Proceedings of the 14th European Conference on Innovation and Entrepreneurship*, University of Peloponnese, Kalamata, Greece.

Adams, R., Jeanrenaud, S., Bessant, J., Denyer, D., Overy, P. (2015). Sustainability-oriented innovation: A systematic review. *International Journal of Management Reviews*, 18, 180–205.

Afuah, A. (2004). *Business Models – A Strategic Management Approach*. McGraw-Hill/Irwin, New York.

Afuah, A. and Tucci, C.L. (2001). *Internet Business Models and Strategies*. McGraw-Hill/Irwin, New York.

Ahvenniemi, H., Huovila, A., Pinto-Seppä, I., Airaksinen, M. (2017). What are the differences between sustainable and smart cities? *Cities*, 60, 234–245.

Albino, V., Berardi, U., Dangelico, R. (2015). Smart cities: Definitions, dimensions, performance and initiatives. *Journal of Urban Technology*, 22(1), 3–21.

Alt, R. and Zimmerman, H.D. (2001). Introduction to special section on business models. *Electronic Markets*, 11(1), 3–9.

Amit, R. and Zott, C. (2001). Value creation in E-business. *Strategic Management Journal*, 22, 493–520.

Amit, R. and Zott, C. (2012). Creating value through business model innovation. *MIT Sloan Management Review*, 53, 41–49.

Andries, P. and Debackere, K. (2013). Business model innovation: Propositions on the appropriateness of different learning approaches. *Creativity & Innovation Management*, 22, 337–358.

Antoine, A.L.C. (2017). Performance-based project delivery selection for highway and construction projects. PhD Thesis, University of Colorado, Boulder.

Applegate, L. (2001). E-business models: Making sense of the internet business landscape. In *Information Technology and the Future Enterprise, New Models for Managers*, Dickson, G. and De Sanctis, G. (eds). Prentice Hall, Upper Saddle River.

Arnkil, R., Järvensivu, A., Koski, P., Piirainen, T. (2010). Exploring quadruple helix: Outlining user-oriented innovation models. University of Tampere, Institute of Social Research, Work Research Centre, Tampere.

Aspara, J., Hietanen, J., Tikkanen, H. (2010). Business model innovation vs. replication: Financial performance implications of strategic emphases. *Journal of Strategic Marketing*, 18, 39–56.

Aspara, J., Lamberg, J.-A., Laukia, A., Tikkanen, H. (2013). Corporate business model transformation and interorganisational cognition: The case of Nokia. *Long Range Planning*, 46, 459–474.

Baden-Fuller, C. and Haefliger, S. (2013). Business models and technological innovation. *Long Range Planning*, 46(6), 419–426.

Baden-Fuller, C. and Morgan, M.S. (2010). Business models as models. *Long Range Planning*, 43(2), 156–171.

Baden-Fuller, C. and Mangematin, V. (2013). Business models: A challenging agenda. *Strategic Organization*, 11(4), 418–427.

Baden-Fuller, C., Giudici, A., Haefliger, S., Morgan, M. (2017). Business models and value. *Academy of Management Proceedings*, 2017(1), doi:10.5465/AMBPP.2017.90 (abbreviated version of this paper).

Bagchi, S. and Tulskie, B. (2000). E-business models: Integrating learning from strategy development experiences and empirical research. *20th Annual International Conference of the Strategic Management Society*, Vancouver, October, 15–18.

Baker, S., Kousis, M., Richardson, D., Young, S. (1997). *The Politics of Sustainable Development: Theory, Policy and Practice within the European Union*. Routledge, London.

Bakici, T., Almirall, E., Wareham, J. (2012). A smart city initiative: The case of Barcelona. *Journal of the Knowledge Economy*, 2(1), 1–14.

Barlow, J. (2000). Innovation and learning in complex offshore construction projects. *Research Policy*, 29(7–8), 973–989.

Barrett, P., Abbott, C., Ruddock, L., Sexton, M. (2007). Hidden innovation in construction and property sectors. *RICS Research Paper Series*, 7(20), RICS, London.

Batty, M., Axhausen, K.W., Giannotti, F., Pozdoukhov, A., Bazzani, A., Wachowicz, M. (2012). Smart cities of the future. *The European Physical Journal*, 214, 481–518.

Behzadan, A.H. and Kamat, V.R. (2005). Visualization of construction graphics in outdoor augmented reality. *Proceedings of the 37th Conference on Winter Simulation*, 1914–1920.

Berglund, H. and Sandström, C. (2013). Business model innovation from an open systems perspective: Structural challenges and managerial solutions. *International Journal of Product Development*, 18, 274–285.

Bettencourt, L.M.A., Lobo, J., Helbing, D., Kuhnert, C., West, G.B. (2007). Growth, innovation, scaling, and the pace of life in cities. *Proceedings of the National Academy of Sciences*, 104(17), 7301–7306.

Betz, F. (2002). Strategic business models. *Engineering Management Journal*, 14(1), 21–27.

Bibri, S.E. and Krogstie, J. (2017). Smart sustainable cities of the future: An extensive interdisciplinary literature review. *Sustainable Cities and Society*, 31, 183–212.

Blayse, A.M. and Manley, K. (2004). Key influences on construction innovation. *Construction Innovation*, 4, 143–154.

Bocken, N.M. (2015). Sustainable venture capital – Catalyst for sustainable start-up success? *Journal of Cleaner Production*, 8, 647–658.

Bogdan, R. and Biklen, S.K. (2006). *Qualitative Research for Education: An Introduction to Theories and Methods*, 5th edition. Pearson, London.

Bonaccorsi, A., Giannangeli, S., Rossi, C. (2006). Entry strategies under competing standards: Hybrid business models in the open source software industry. *Management Science*, 52, 1085–1098.

Boons, F. and Lüdeke-Freund, F. (2013). Business models for sustainable innovation: State-of-the-art and steps towards a research agenda. *Journal of Cleaner Production*, 45, 9–19.

Bouskela, M., Casseb, M., Bassi, S., Facchina, M. (2016). The road toward smart cities: Migrating from traditional city management to the smart city. *Inter-American Development Bank*, July [Online]. Available at: https://publications.iadb.org/en/road-toward-smart-cities-migrating-traditional-city-management-smart-city.

Brousseau, E. and Penard, T. (2006). The economics of digital business models: A framework for analyzing the economics of platforms. *Review of Network Economics*, 6(2), 81–110.

Brusoni, S., Precipe, A., Salter, A. (1998). Mapping and measuring innovation in project-based firms. CoPS Working Paper, Brighton.

Bucherer, E., Eisert, U., Gassmann, O. (2012). Towards systematic business model innovation: Lessons from product innovation management. *Creativity & Innovation Management*, 21, 183–198.

Calia, R.C., Guerrini, F.M., Moura, G.L. (2007). Innovation networks: From technological development to business model reconfiguration. *Technovation*, 27, 426–432.

Campbell, I., Bourell, D., Gibson, I. (2012). Additive manufacturing: Rapid prototyping comes of age. *Rapid Prototyping Journal*, 18(4), 255–258.

Caragliu, A. and Del Bo, C. (2016). Do smart cities invest in smarter policies? Learning from the past, planning for the future. *Social Science Computer Review*, 34(6), 657–672.

Caragliu, A., Del Bo, C., Nijkamp, P. (2011). Smart cities in Europe. *Journal of Urban Technology*, 18(2), 65–82.

Carassus, J. (2004). Construction sector system, innovation and services. *CIB Congress Proceedings*, Toronto, Canada [Online]. Available at: http://jc-desh.cstb.fr/file/rub49_doc46_1.pdf.

Casadesus-Masanell, R. and Ricart, J.E. (2007). Competing through business models. Working Paper WP 713, November, IESE Business School, University of Navarra, Barcelona.

Casadesus-Masanell, R. and Ricart, J.E. (2010). From strategy to business models and onto tactics. *Long Range Planning*, 43(2–3), 195–215.

Casadesus-Masanell, R. and Ricart, J.E. (2011). How to design a winning business model. *Harvard Buisness Review*, 89(1–2), 100–107.

Casadesus-Masanell, R. and Tarzijan, J. (2012). When one business model isn't enough. *Harvard Business Review*, 90(1–2).

Casadesus-Masanell, R. and Zhu, F. (2013). Business model innovation and competitive imitation: The case of sponsor-based business models. *Strategic Management Journal*, 34, 464–482.

Chesbrough, H.W. (2003). *Open Innovation: The New Imperative for Creating and Profiting from Technology*. Harvard Business School Press, Boston.

Chesbrough, H.W. (2007). Business model innovation: It's not just about technology anymore. *Strategy and Leadership*, 35, 12–17.

Chesbrough, H.W. (2010) Business model innovation: Opportunities and barriers. *Long Range*, 43(2–3), 354–363.

Chesbrough, H.W. and Rosenbloom, R.S. (2002). The role of the business model in capturing value from innovation: Evidence from Xerox Corporation's technology spinoff companies. *Industrial and Corporate Change*, 11, 533–534.

Christensen, C.M., Baumann, H., Ruggles, R., Sadtler, T.M. (2006). Disruptive innovation for social change. *Harvard Business Review*, 84(12), 94–101.

Chui, M., Lund, S., Madgavkar, A., Mischke, J., Ramaswamy, S. (2018). Smart cities: Digital solutions for a more livable future. *McKinsey Global Institute*, June.

Chun, Y.Y. and Lee, K.M. (2013). Life cycle-based generic business strategies for sustainable business models. *Journal of Sustainable Development*, 6(8), 1.

Cloutier, C. (2005). Processus d'élaboration d'un projet de construction. Guide, Direction des immobilisations, Ministère de la Culture et des Communications du Québec (MCC), Quebec.

Cohen, B. (2012). What exactly is a smart city? *Fast Company* [Online]. Available at: https://www.fastcodesign.com/1680538/what-exactly-is-a-smart-city.

Cretu, G.L. (2012). Smart cities design using event-driven paradigm and semantic web. *Informatica Economica*, 16(4), 57–67.

Cronin, B. and Davenport, E. (1991). The compound eye/I: An introduction to social intelligence. *Social Intelligence*, 1(1), 1–6.

Cucculelli, M. and Bettinelli, C. (2015). Business models, intangibles and firm performance: Evidence on corporate entrepreneurship from Italian manufacturing SMEs. *Small Business Economics*, 45, 329–350.

Da Rocha, C.G., Formoso, C., Fazenda, P., Koskela, L., Tezel, A. (2012). Design science research in lean construction: Process and outcomes. In *IGLC-20*, *Proceedings,* San Diego.

Damanpour, F. (1992). Organizational size and innovation. *Organisation Studies*, 13(3), 375–402.

Damanpour, F. (1996). Organizational complexity and innovation: Developing and testing multiple contingency models. *Management Science*, 42, 693–716.

Dameri, R.P. (2013). Searching for smart city definition: A comprehensive proposal. *International Journal of Computers & Technology*, 11(2), 2544–2551.

DaSilva, C.M. and Trkman, P. (2014). Business model: What it is and what it is not. *Long Range Planning*, 47(6), 379–389.

Deakin, M. (2012). *Smart Cities: Governing, Modelling and Analyzing the Transition.* Routledge, London.

Dedijer, S. and Jéquier, N. (1987). *Intelligence for Economic Development: An Inquiry into the Role of the Knowledge Industry*. Berg, Oxford.

Deimler, M. and Kachaner, N. (2020). Business model innovation. *A BCG Publication* [Online]. Available at: https://www.bcg.com/capabilities/strategy/business-model-innovation.aspx.

Demil, B. and Lecocq, X. (2008). (Re)penser le développement des organisations. Les apports du modèle économique. *Revue Française de Gestion*, 181, 113–122.

Demil, B. and Lecocq, X. (2010). Business model evolution: In search of dynamic consistency. *Long Range Planning*, 43, 227–246.

Demil, B., Lecocq, X., Ricart, J.E., Zott, C. (2015). Introduction to the SEJ special issue on business models: Business models within the domain of strategic entrepreneurship. *Strategic Entrepreneurship Journal*, 9(1), 1–11.

Doganova, L. and Eyquem-Renault, M. (2009). What do business models do? Innovation devices in technology entrepreneurship. *Research Policy*, 38, 1559–1570.

Doz, Y.L. and Kosonen, M. (2010). Embedding strategic agility: A leadership agenda for accelerating business model renewal. *Long Range Planning*, 43, 370–382.

Egbu, C. (2008). Clients' roles and contributions to innovations in the construction industry: When giants learn to dance. In *Clients Driving Innovation*, Brandon, P. and Lu, S.L. (eds). Blackwell Publishing, Chichester.

Eisenmann, T. (2002). *Internet Business Models, Text and Cases*. McGraw-Hill/Irwin, New York.

Elkington, J. (1997). *Cannibals with Forks: The Triple Bottom Line of 21st Century Business*. Capstone, Oxford.

Eremia, M., Toma, L., Manoloiu, A., Sanduleac, M. (2017). The smart city concept in the 21st century. *Procedia Engineering*, 181, 12–19.

Etezadzadeh, C. (2016). *Smart City – Future City*? Springer Vieweg, Wiesbaden, Germany.

Evans, J.D. and Johnson, R.O. (2013). Tools for managing early-stage business model innovation. *Research Technology Management*, 56, 52–56.

Ferrero, F., Perboli, G., Vesco, A., Musso, S., Pacifici, A. (2015). Car-sharing services – Part B: Business & service models. Paper, CIRRELT, Montreal.

Flyvbjerg, B. (2014). What you should know about megaprojects and why: An overview. *Project Management Journal*, 45(2), 6–19.

Flyvbjerg, B. and Holm, M.S. (2002). Underestimating cost in public works projects. *Journal of the American Planning Association*, 68, 279–295.

Flyvbjerg, B., Bruzelius, N., Rothengatter, W. (2003). *Megaprojects and Risks*. Cambridge University Press, Cambridge.

Foss, N.J. and Saebi, T. (2017). Fifteen years of research on business model innovation. *Journal of Management*, 43(1), 200–227.

Gambardella, A. and McGahan, A.M. (2010). Business-model innovation: General purpose technologies and their implications for industry structure. *Long Range Planning*, 43, 262–271.

Gann, D.M. and Salter, A. (2000). Innovation in project-based, service-enhanced firms: The construction of complex products and systems. *Research Policy*, 29(7–8), 955–972.

Gassman, O., Frankenberger, K., Csik, M. (2014). *The Business Model Navigator*. FT Publishing, Upper Saddle River.

Gawer, A. (2014). Bridging differing perspectives on technological platforms: Toward an integrative framework. *Research Policy*, 43, 1239–1249.

Gawer, A. and Cusumano, M.A. (2014). Industry platforms and ecosystem innovation. *Journal of Product Innovation Management*, 31(3), 417–433.

Georg, S., De Linde, G.G., Pinheiro-croisel, R., Eco, F.A. (2012). Eco-districts and sustainable cities – Institutionalization through experimentation. *Academy of Management Meeting*, June 2011, 1–38.

Giesen, E., Berman, S.J., Bell, R., Blitz, A. (2007). Three ways to successfully innovate your business model. *Strategy & Leadership*, 35(6), 27–33.

Giffinger, R., Fertner, C., Kramar, H., Kalasek, R., Pichler-Milanovic, N., Meijers, E. (2007). *Smart Cities: Ranking of European Medium-Sized Cities*. Center of Regional Science, Vienna.

Gordijn, J. and Akkermans, H. (2001). Designing and evaluating e-business models. *Intelligent e-Business*, July–August, 11–17.

Gransberg, D.D. and Shane, J.S. (2015). Defining best value for construction manager/general contractor projects: The CMGC learning curve. *Journal of Management in Engineering*, 31(4), 04014060 [Online]. Available at: https://doi.org/10.1061/(ASCE)ME.1943-5479.0000275.

Greenfield, A. (2006). *Everyware: The Dawning Age of Ubiquitous Computing*. New Riders, Boston.

Gronroos, C. and Voima, P. (2013). Critical service logic: Making sense of value creation and co-creation. *Journal of Academy of Marketing*, 41, 133–150.

Guerrero-Pérez, A.D., Huerta, A., Gonzalez, F., Lopez, D. (2013). Network architecture based on virtualized networks for smart cities. White Papers from the Smart Cities of the Future Kickoff Event, Guadalajara, Mexico, October.

Günzel, F. and Holm, A.B. (2013). One size does not fit all: Understanding the front-end and backend of business model innovation. *International Journal of Innovation Management*, 17(1), 1340002.

Hall, R.E. (2000). The vision of a smart city. *Proceedings of the 2nd International Life Extension Technology Workshop*, Brookhaven National Laboratory, New York, 1–6.

Hanna, K.N. (2009). *Enabling Enterprise Transformation. Innovation, Technology, and Knowledge Management Series*. Springer, University of Maryland, Bethesda.

Hansen, M.T. and Birkinshaw, J. (2007). The innovation value chain. *Harvard Business Review*, 85(6), 121–130.

Harrison, C. and Donnelly, I. (2011). A theory of smart cities. *Proceedings of the 55th Annual Meeting ISSS*, 55(1), 1–15.

Harrison, C., Eckman, B., Hamilton, R., Hartswick, P., Kalagnanam, J., Paraszczak, J. (2010). Foundations for smarter cities. *IBM Journal of Research and Development*, 54(4), 1–16.

Hart, S.L. and Milstein, M. (2003). Creating sustainable value. *Academy of Management Executive*, 17. 10.5465/AME.2003.10025194.

Hernandez-Munoz, J. and Munoz, L. (2013). The SmartSantander Project. In *The Future Internet Assembly. Lecture Notes in Computer Science*, Galis, A. and Gavras, A. (eds). Springer, Berlin, Heidelberg.

Höjer, M. and Wangel, J. (2015). Smart sustainable cities: Definition and challenges. In *ICT Innovations for Sustainability: Advances in Intelligent Systems and Computing*, Hilty, L. and Aebischer, B. (eds). Springer, Heidelberg.

Iivari, J. (2007). A paradigmatic analysis of information systems as a design science. *Scandanavian Journal of Information Systems*, 19(2), 39–63.

ITU (2019). Shaping smarter and more sustainable cities: Striving for sustainable development goals. Report, ITU-T Focus Group on Smart Sustainable Cities, Geneva.

Johnson, M.W. and Suskewicz, J. (2009). How to jump-start the clean tech economy. *Harvard Business Review*, 87(11), 52–60.

Johnson, M.W., Christensen, C.M., Kagermann, H. (2008). Reinventing your business model. *Harvard Business Review*, 86(12), 50–59.

de Jong, M., Joss, S., Schraven, D. (2015). Sustainable-smart-resilient-low-carbon-eco-knowledge cities: Making sense of a multitude of concepts promoting sustainable urbanization. *Journal of Cleaner Production*, 109, 25–38.

Ke, H., Cui, Z., Govindan, K., Zavadskas, A.K. (2015). The impact of contractual governance and trust on EPC projects in construction supply chain performance. *Engineering Economics*, 26(4), 349–363.

Khanagha, S., Volberda, H., Oshri, I. (2014). Business model renewal and ambidexterity: Structural alteration and strategy formation process during transition to a Cloud business model. *R&D Management*, 44, 322–340.

Khanna, T., Oberholzer-Gee, F., Dessain, V.M., Jensen, A.D., Sjoman, A. (2007). Metro International S.A. *Harvard Business School*, Case 708–429, September.

Kiani, B., Gholamian, M.R., Hamzehei, A., Hosseini, S.H. (2009). Using causal loop diagram to achieve a better understanding of e-business models. *International Journal of Electronic Business Management*, 7(3), 159–167.

Kitchin, R. (2014). The real-time city? Big data and smart urbanism. *GeoJournal*, 79, 1–14.

Koen, P., Bertels, H.M.J., Elsum, I.R. (2011). The three faces of business model innovation: Challenges for established firms. *Research Technology Management*, 54(3), 52–59.

Kominos, N., Kakderi, C., Brastas, C., Tsarchopoulos, P. (2015). Smart city ontologies: Improving the effectiveness of smart city application. *Journal of Smart Cities*, 1, 1–16, doi: 10.18063/JSC.2015.01.001.

Komleh, R.A. (2018). Theory of value from Smith to Marshall. University of Milan, Milan.

Kordas, O., Nielsen, J., Neij, L. (2015). Strategic innovation program for smart sustainable cities: Application for a strategic innovation program [Online]. Available at: https://smartsustainablecities.se/wpcontent/uploads/2017/05/SSC SIP_final.pdf.

Koskela, (1992). Application of the new production philosophy to construction. Centre for Integrated Facility Engineering, Technical report # 72. Stanford University, Stanford.

Kuechler, W.L. and Vaishnavi, V.K. (2008). The emergence of design research in information systems in North America. *Journal of Design Research*, 7(1), 1–16.

Kumaraswamy, A., Garud, R., Ansari, S. (2018). Perspectives on disruptive innovations. *Journal of Management Studies* (*Special Issue: Managing in the Age of Disruptions*), doi:10.1111/Joms.12399.

Lambert, S.C. and Davidson, R.A. (2013). Applications of the business model in studies of enterprise success, innovation and classification: An analysis of empirical research from 1996 to 2010. *European Management Journal*, 31, 668–681.

Lansley, P. (1996). Innovation: The role of research, education and practice. In *Construction Papers*, Harlow, P. (ed.). CIOB, Ascot.

Laszczuk, A., Garreau, L., de Montmorillon, B. (2017). Understanding emergence in business model development: How companies interact with stakeholders to deal with environmental ambiguity. Dauphine Recherches en Management (DRM), HAL Id: hal-01787276.

Lazaroiu, G.C. and Roscia, M. (2012). Definition methodology for the smart cities model. *Energy*, 47(1), 326–332.

Le Masson, P. and Weil, B. (2009). La domestication de l'innovation dans les entreprises industrielles : l'invention des bureaux d'études. *Les nouveaux régimes de la conception, langages, théories, métiers*, 51–66 [Online]. Available at: papers3://publication/uuid/C6A89146-C894-4B8E-B2F8-8D299A7E0A27%5Cnpapers3://publication/uuid/636D7D4F-325C-4A0D-9FCD-DF9924308DA9.

Lecocq, X., Demil, B., Warnier, V. (2006). Le business model, un outil d'analyse stratégique, *L'expansion management review*, 96–109.

Lee, M. and Stinson, D. (2014). Organizational decision making models: Comparing & contrasting to the Stinson wellness model. *European Journal of Management*, 14(13), 13–28.

Lehdonvirta, V., Wilska, T.A., Johnson, M. (2009). Virtual consumerism: Case Habbo Hotel, Information. *Communication & Society*, 12(7), 1059–1079.

Lim, S., Buswell, R.A., Le, T.T., Austin, S.A., Gibb, A.G.F., Thorpe, T. (2012). Developments in construction-scale additive manufacturing processes. *Automation in Construction*, 21(1), 262–268 [Online]. Available at: https://hdl.handle.net/2134/9176.

Lindgart, Z., Reeves, M., Stalk, G., Deimier, S. (2009). *Business Model Innovation: When the Game Gets Tough, Change the Game.* The Boston Consulting Group, Boston.

Lozano, R. (2013). Are companies planning their organisational changes for corporate sustainability? An Analysis of three case studies on resistance to change and their strategies to overcome it. *Corporate Social Responsibility and Environmental Management*, 20(5), 275–295.

Lubell, M. (2015). Collaborative partnerships in complex institutional systems. *Current Opinion in Environmental Sustainability*, 12, 41–47.

Lucking-Reiley, D. and Spulber, D.F. (2001). Business-to-business electronic commerce. *Journal of Economic Perspectives*, 15(1), 55–68.

Lukka, K. (2003). The constructive research approach. In *Case Study Research in Logistics*, Ojala, L. and Hilmola, O.-P. (eds). Publications of the Turku School of Economics and Business Administration, Turku.

Magretta, J. (2002). Why business models matter. *Harvard Business Review*, 80(5), 86–92.

Mahadevan, B. (2000). Business models for internet-based e-commerce: An anatomy. *California Management Review*, 42(4), 55–69.

Marceau, J., Houghton, J., Toner, P., Manley, K., Gerasimou, E., Cook, N. (1999). Mapping the building and construction product system in Australia. White paper, Commonwealth Department of Industry, Science and Resources, Sydney.

March, S. and Smith, G. (1995). Design and natural science research on information technology. *Decision Support Systems*, 15(4), 251–266.

Marjavonic, S., Fry, C., Chataway, J. (2012). Crowdsourcing based business models: In search of evidence for innovation 2.0. *Science and Public Policy*, 39(10).

Markides, C. (2006). Disruptive innovation: In need of better theory. *Journal of Product Innovation Management*, 23, 19–25.

Markides, C. (2013). Business model innovation: What can ambidexterity literature teach us? *Academy of Management Perspectives*, 27, 313–323.

Marsal-Llacuna, M.-L., Colomer-Llinàs, J., Meléndez-Frigola, J. (2015). Lessons in urban monitoring taken from sustainable and livable cities to better address the smart city initiatives. *Technological Forecasting and Social Change*, 90(B), 611–622.

Massa, L. and Tucci, C.L. (2014). Business model innovation. In *The Oxford Handbook of Innovation Management*, Dodgson, M., Gann, D.M., Phillips, N. (eds). Oxford University Press, Oxford.

Massa, L., Tucci, C.L., Afuah, A. (2017). A critical assessment of business model research. *Academy of Management Annals*, 11(1), 73–104.

Matzler, K., Bailom, F., den Eichen S.F., Kohler, T. (2013). Business model innovation: Coffee triumphs for Nespresso. *Journal of Business Strategy*, 34, 30–37.

McGrath, R.G. (2010). Business models: A discovery driven approach. *Long Range Planning*, 43, 247–261.

Meadows, D. (1999). Indicators and information systems for sustainable development. In *The Earthscan Reader in Sustainable Cities*, Satterthwaite, D. (ed.). Earthscan, London.

Meijer, A.J., Gil-Garcia, J.R., Bolívar, M.P.R. (2016). Smart city research. *Social Science Computer Review*, 34(6), 647–656.

Mitchell, D. and Coles, C. (2003). The ultimate competitive advantage of continuing business model innovation. *Journal of Business Strategy*, 24, 15–21.

Mitchell, D. and Coles, C. (2004). Business model innovation breakthrough moves. *Journal of Business Strategy*, 25, 16–26.

Moir, E., Moonem, T., Clark, G. (2014). *What are Future Cities? Origins, Meaning and Uses.* Government Office of Science, London.

Morris, M., Schindehutte, M., Allen, J. (2005). The entrepreneur's business model: Toward a unified perspective. *Journal of Business Research*, 58(6), 726–735.

Morris, M., Schindehutte, M., Richardson, J., Allen, J. (2006). Is the business model a useful strategic concept? Conceptual, theoretical, and empirical insights. *Journal of Small Business Strategy*, 17(1), 27–49.

Najam, A. (2005). Developing countries and global environmental governance: From contestation to participation to engagement. *International Environmental Agreements*, 5, 303–321.

Nam, T. and Pardo, T.A. (2011). Conceptualizing the smart city with dimensions of technology, people, and institutions. *Proceedings of the 12th Conference on Digital Government Research*, Association for Computing Machinery, College Park, 282–288.

Neirotti, P., De Marco, A., Cagliano, A.C., Mangano, G., Scorrano, F. (2014). Current trends in smart city initiatives: Some stylized facts. *Cities*, 38, 25–36.

Normann, R. and Ramirez, R. (1993). From value chain to value constellation: Designing interactive strategy. *Harvard Business Review*, 71(4), 65–77.

Nunamaker, J.F., Chen, M., Purdin, T.D.M. (1990). Systems development in information systems research. *Journal of Management Information Systems*, 7(3), 89–106.

Osterwalder, A. (2004). The business model ontology: A proposition in a design science approach. Doctoral Thesis, University of Lausanne, Switzerland.

Osterwalder, A. and Pigneur, Y. (2010). Business model generation: A handbook for visionaries. *Game Changers, and Challengers*, John Wiley & Sons Inc., Hoboken.

Osterwalder, A., Pigneur, Y., Tucci, C.L. (2005). Clarifying business models: Origins, present, and future of the concept. *Communications of the Association for Informations Systems*, 16, 1–25.

Pateli, A.G. and Giaglis, G.M. (2005). Technology innovation-induced business model change: A contingency approach. *Journal of Organisational Change Management*, 18, 167–183.

Pauwels, K. and Weiss, A. (2008). Moving from free to fee: How online firms market to change their business model successfully. *Journal of Marketing*, 72(3), 14–31.

Pavie, X., Hsu, E., Justus, T.R., Tapia, R.O. (2013). How to define and analyze business model innovation in service. Working paper [Online]. Available at: https://hal-essec.archives.ouvertes.fr/hal-00921420.

Pekuri, A., Herrala, M., Aapaoja, A., Haapasalo, H. (2012). Applying lean in construction – Cornerstones for implementation. In Tommelein, I.D. and Pasquire, C.L. (eds). *20th Annual Conference of the International Group for Lean Construction*, San Diego.

Peppard, J. and Rylander, A. (2006). From value chain to value network: Insights for mobile operators. *European Management Journal*, 24(2), 128–141.

Pepper, D. (1998). Sustainable development and ecological modernization: A radical homocentric perspective. *Sustainable Development*, 6, 1–7.

Pfeffers, K., Tuuanen, T., Rothenberger, M.A., Chatterjee, S. (2007). A design science research methodology for information systems research. *Journal of Management Information Systems*, 24(3) 45–77.

Phillips, R. (1997) Innovation and firm performance in Australian manufacturing. Staff Research Paper, Industry Commission, Canberra.

Pigneur, Y., Joyce, A., Paquin, R. (2015). The triple layered business model canvas: A tool to design more sustainable business models. *Journal of Cleaner Production*, 135, 1474–1486.

Pinch, T.J. and Bijker, W.E. (1984). The social construction of facts and artefacts: Or how sociology of science and the sociology of technology might benefit each other. *Social Studies of Science*, 14(3), 399–441.

Porter, M.E. (1985). *Competitive Advantage: Creating and Sustaining Superior Performance*. The Free Press, New York.

Porter, M.E. (2001). Strategy and the internet. *Harvard Business Review*, 79(3), 62–78.

Porter, M.E. and Kramer, M. (2011). Creating shared value. *Harvard Business Review*, 89(1), 62–77.

Prahalad, C.K. and Hart, S. (2004). *The Fortune at the Bottom of the Pyramid*. Wharton School Publishing, Upper Saddle River, NJ.

Prizzia, R. (2007). Sustainable development in an international perspective. In *Handbook of Globalization and the Environment*, Thai, K.V., Rahm, D., Coggburn, J.D. (eds). CRC Press, Boca Raton.

Ramanathan, C., Sambu, N., Idrus, A. (2012). Construction delays causing risks on time and cost – A critical review. *Australasian Journal of Construction Economics and Building*, 12, 37–57.

Rappa, M. (2001). Business models on the web. Managing the digital enterprise [Online]. Available at: digitalenterprise.org/models/models.html [Accessed December 2009].

Rittel, H.W.J. and Webber, M.M. (1973). Dilemmas in a general theory of planning. *Policy Sciences*, 4, 155–169.

Rivkin, J.W. (2000). Dogfight over Europe: Ryanair (A). *Harvard Business School*, Case 700–115, June (Revised November 2007).

Rochet, C. (2014). Les villes intelligentes, enjeux et stratégies pour de nouveaux marchés. Étude réalisée avec le concours de CERGAM, CESAMES et du service de coordination à l'intelligence économique. Conference paper, Ministère français des finances et des comptes publics, Paris.

Rochet, J.-C. and Tirole, J. (2006). Two-sided markets: A progress report. *The RAND Journal of Economics*, 37(3), 645–667.

Rode, P. and Burdett, R. (2011). Cities: Investing in energy and resource efficiency. In *Towards a Green Economy: Pathways to Sustainable Development and Poverty Eradication*, United Nations Environment (ed.). UNEP, Nairobi.

Rodolfo, A. and Sy-Changco, J. (2015). The phenomenon of sachet marketing: Lessons to be learned from the Philippines. *Advancing Knowledge Development in Marketing*, 18, 5–15.

Rogers, R. (1998). *Cities for a Small Planet*. Westview Press, Boulder.

Rosati, U. and Conti, S. (2016). What is a smart city project? An urban model or a corporate business plan? *Procedia – Social and Behavioral Sciences*, 223, 968–973.

Rowland, W. (1973). *The Plot to Save the World*. Clarke Irwin & Co, Toronto.

Sabatier, V., Craig-Kennard, A., Mangematin, V. (2012). When technological discontinuities and disruptive business models challenge dominant industry logics: Insights from the drugs industry. *Technological Forecasting and Social Change*, 79, 946–962.

Sánchez, P. and Ricart, J.E. (2010). Business model innovation and sources of value creation in low-income markets. *European Management Review*, 7, 138–154.

Sang Keon, L., Heeseo, R.K., HeeAh, C., Jongbok, K., Donju, L. (2016). International case studies of smart cities: Songdo, Republic of Korea. Institutions for Development Sector, Fiscal & Municipal Management Division, Discussion Paper IDB-DP-463, June.

Santoni, G. and Taglioni, D. (2015). *The Age of Global Value Chains*. CEPR Press, London.

Santos, J., Spector, B., Van der Heyden, L. (2009). Toward a theory of business model innovation within incumbent firms. Working Paper No. 16, INSEAD, Fontainebleau, France.

Schaffers, H., Komninos, N., Pallot, M., Trousse, B., Nilsson, M., Oliveira, A. (2011). Smart cities and the future internet: Towards cooperation frameworks for open innovation. In *The Future Internet*, Domingue, J. (ed.). Springer, Heidelberg.

Schiavi, G. and Behr, A. (2018). Emerging technologies and new business models: A review on disruptive business models. *Innovation & Management Review*, 15(4), 338–355.

Schneider, S. and Spieth, P. (2013). Business model innovation: Towards an integrated future research agenda. *International Journal of Innovation Management*, 17, 134–156.

Seaden, G. and Manseau, A. (2001) Public policy and construction innovation. *Building Research and Information*, 29(3), 182–196.

Sherman, W.R. (2012). The triple bottom line: The reporting of doing well & doing good. *Journal of Applied Business Research (JABR)*, 28(4), 673–682.

Shrivastava, P. and Statler, M. (2012). *Learning from the Global Financial Crisis: Creatively, Reliably, and Sustainably*. Stanford University Press, Palo Alto.

Silk, K. and Appleby, J.W. (2010). *Open Data, Open City*. Martin Prosperity Institute, Toronto.

Simard, J. (2015). La ville intelligente comme vecteur pour le développement durable : le cas de la ville de montréal. Thesis, Université de Sherbrooke, Quebec.

SIP-SSC. (2015). Strategic innovation agenda for smart sustainable cities [Online]. Available at: https://www.diva-portal.org/smash/get/diva2:812341/FULLTEXT01.pdf.

Slaughter, S.E. (1998). Models of construction innovation. *Journal of Construction Engineering and Management*, 124(3), 226–231.

Slaughter, S.E. (2000). Implementation of construction innovations. *Building Research and Information*, 28(1), 2–17.

Smart Cities India (2015). Smarter solutions for a better tomorrow [Online]. Available at: https://eusmartcities.eu/sites/all/files/events/uploads/Smart%20Cities%20India%202015%20Brochure_0.pdf.

Söderberg, P.J. (2019). Nespresso case study. *Entrepreneurship & Innovation*, 10.

Souto, J.E. (2015). Business model innovation and business concept innovation as the context of incremental innovation and radical innovation. *Tourism Management*, December.

Spieth, P., Schneckenberg, D., Ricart, J.E. (2014). Business model innovation – State of the art and future challenges for the field. *R&D Management*, 44(3), 237–247.

Stabell, B. and Fjeldstad, O. (1998). Configuring value for competitive advantage: On chains, shops, and networks. *Strategic Management Journal*, 9(5), 413–437 [Online]. Available at: http://www.jstor.org/stable/3094221.

Stewart, D.W. and Zhao, Q. (2000). Internet marketing, business models and public policy. *Journal of Public Policy and Marketing*, 19, 287–296.

Stonehouse, G. and Snowdon, B. (2007). Competitive advantage revisited: Michael Porter on strategy and competitiveness. *Journal of Management Inquiry*, 16(3), 256–273.

Tapscott, D., Ticoll, D., Lowy, A. (2000). *Digital Capital: Harnessing the Power of Business Webs*. Harvard Business School Press, Boston.

Teece, D.J. (2010). Business model, business strategy and innovation. *Long Range Planning*, 43(2–3), 172–194.

Termeer, C.J. and Bruinsma, A. (2016). ICT-enabled boundary spanning arrangements in collaborative sustainability governance. *Current Opinion in Environmental Sustainability*, 18, 91–98.

Thompson, J.D. and MacMillan, I.C. (2010). Business models: Creating new markets and societal wealth. *Long Range Planning*, 43, 291–307.

Timmers, P. (1998). Business models for electronic markets. *Electronic Markets*, 8(2), 3–8.

Tobergate, D.R. and Curtis, S. (2014). Mapping smart cities in the EU. *Journal of Chemical Information and Modeling*, 53(9), 1689–1699.

Townsend, A. (2013). *Smart Cities: Big Data, Civic Hackers, and the Quest for a New Utopia*. W.W. Norton & Co, New York.

Turok, I. (2014). Cities as drivers of development. In *Urbanization and Socioeconomic Development in Africa: Challenges and Opportunities*, Kayizza-Mugerwa, S., Shimeles, A., Yameogo, N.D. (eds). Routledge, London.

UN-Habitat, (2016). *World Cities Report 2016: Urbanization and Development: Emerging Futures*. UN-Habitat, Nairobi.

UNEP (2002). *Melbourne Principles for Sustainable Cities*. UNEP, Nairobi.

United Nations (2013). World economic and social survey: Sustainable development challenges. United Nations, New York [Online]. Available at: https://www.un.org/en/development/desa/policy/wess/wess_current/wess2013/WESS2013.pdf.

Vargo, S.L. and Lusch, R. (2004). Evolving to a new dominant logic. *Journal of Marketing*, 68(1), 1–17.

Virtanen, J.P., Hyyppä, H., Kurkela, M., Vaaja, M., Alho, P., Hyyppä, J. (2014). Rapid prototyping – A tool for presenting 3-dimensional digital models produced by terrestrial laser scanning. *ISPRS International Journal of Geo-Information*, 3(3), 871–890.

Voelpel, S.C., Leibold, M., Tekie, E.B. (2004). The wheel of business model reinvention: How to reshape your business model to leapfrog competitors. *Journal of Change Management*, 4, 259–276.

Vrijhoef, R. and Koskela, L. (2000). The four roles of supply chain management in construction. *European Journal of Purchasing & Supply Management*, 6(3), 169–178.

Washburn, D., Sindhu, U., Balaouras, S., Dines, R.A., Hayes, N.M., Nelson, L.E. (2010). *Helping CIOs Understand Smart City Initiatives: Defining the Smart City, Its Drivers, and the Role of the CIO*. Forrester Research, Cambridge, MA.

WCED (1987). *Our Common Future: Report of the World Commission on Environment and Development*. Oxford University Press, Oxford.

Weber, R. (1987). Toward a theory of artifacts: A paradigmatic base for information systems research. *Journal of Information Systems*, 1(2), 3–19.

Wei, Z., Yang, D., Sun, B., Gu, M. (2014). The fit between technological innovation and business model design for firm growth: Evidence from China. *R&D Management*, 44, 288–305.

Weill, P. and Vitale, M. (2001). *Place to Space*. Harvard Business School Press, Boston.

Weill, P. and Woerner, S.L. (2015). Thriving in an increasingly digital ecosystem. *MIT Sloan Management Review*, 56(4), 27–34.

Wirtz, B.W., Pistoia, A., Ullrich, S., Gottel, V. (2015). Business models: Origin, development and future research. *Long Range Planning*, 49, 36–54.

Yang, J., Yang, C., Kao, C. (2010). Evaluating schedule delay causes for private participating public construction works under the build-operate-transfer model. *International Journal of Project Management*, 28, 569–579.

Yip, G. (2004). Using strategy to change your business model. *Business Strategy Review*, 15(2), 17–24.

Yunus, M., Moingeon, B., Lehmann-Ortega, L. (2010). Building social business models: Lessons from the Grameen experience. *Long Range Planning*, 43, 308–325.

Zott, C. and Amit, R. (2008). The fit between product market strategy and business model: Implications for firm performance. *Strategic Management Journal*, 29(1), 1–26.

Zott, C. and Amit, R. (2010). Business model design: An activity system perspective. *Long Range Planning*, 43(2–3), 216–226.

Zott, C., Amit, R., Massa, L. (2011). The business model: Recent developments and future research. *Journal of Management*, 37(4), 1019–1042.

Index

B

C, E

F, G, I

M, O

P, Q, R, S

T, U, V, W

Other titles from

iSTE

in

Information Systems, Web and Pervasive Computing

2021

BEN REBAH Hassen, BOUKTHIR Hafedh, CHÉDEBOIS Antoine
Website Design and Development with HTML5 and CSS3

EL ASSAD Safwan, BARBA Dominique
Digital Communications 1: Fundamentals and Techniques
Digital Communications 2: Directed and Practical Work

GAUDIN Thierry, MAUREL Marie-Christine, POMEROL Jean-Charles
Chance, Calculation and Life

SÉBASTIEN-YVES Laurent
Conflicts, Crimes and Regulations in Cyberspace
(Cybersecurity Set – Volume 2)

LE DEUFF Olivier
Hyperdocumentation (Intellectual Technologies Set – Volume 9)

PÉLISSIER Maud
Cultural Commons in the Digital Ecosystem
(Intellectual Technologies Set – Volume 8)

2020

CLIQUET Gérard, with the collaboration of BARAY Jérôme
Location-Based Marketing: Geomarketing and Geolocation

DE FRÉMINVILLE Marie
Cybersecurity and Decision Makers: Data Security and Digital Trust

GEORGE Éric
Digitalization of Society and Socio-political Issues 2: Digital, Information and Research

HELALI Saida
Systems and Network Infrastructure Integration

LOISEAU Hugo, VENTRE Daniel, ADEN Hartmut
Cybersecurity in Humanities and Social Sciences: A Research Methods Approach (Cybersecurity Set – Volume 1)

SEDKAOUI Soraya, KHELFAOUI Mounia
Sharing Economy and Big Data Analytics

SCHMITT Églantine
Big Data: An Art of Decision Making
(Intellectual Technologies Set – Volume 7)

2019

ALBAN Daniel, EYNAUD Philippe, MALAURENT Julien, RICHET Jean-Loup, VITARI Claudio
Information Systems Management: Governance, Urbanization and Alignment

AUGEY Dominique, with the collaboration of ALCARAZ Marina
Digital Information Ecosystems: Smart Press

BATTON-HUBERT Mireille, DESJARDIN Eric, PINET François
Geographic Data Imperfection 1: From Theory to Applications

BRIQUET-DUHAZÉ Sophie, TURCOTTE Catherine
From Reading-Writing Research to Practice

BROCHARD Luigi, KAMATH Vinod, CORBALAN Julita, HOLLAND Scott, MITTELBACH Walter, OTT Michael
Energy-Efficient Computing and Data Centers

CHAMOUX Jean-Pierre
The Digital Era 2: Political Economy Revisited

COCHARD Gérard-Michel
Introduction to Stochastic Processes and Simulation

DUONG Véronique
SEO Management: Methods and Techniques to Achieve Success

GAUCHEREL Cédric, GOUYON Pierre-Henri, DESSALLES Jean-Louis
Information, The Hidden Side of Life

GEORGE Éric
Digitalization of Society and Socio-political Issues 1: Digital, Communication and Culture

GHLALA Riadh
Analytic SQL in SQL Server 2014/2016

JANIER Mathilde, SAINT-DIZIER Patrick
Argument Mining: Linguistic Foundations

SOURIS Marc
Epidemiology and Geography: Principles, Methods and Tools of Spatial Analysis

TOUNSI Wiem
Cyber-Vigilance and Digital Trust: Cyber Security in the Era of Cloud Computing and IoT

2018

ARDUIN Pierre-Emmanuel
Insider Threats
(Advances in Information Systems Set – Volume 10)

CARMÈS Maryse
Digital Organizations Manufacturing: Scripts, Performativity and Semiopolitics
(Intellectual Technologies Set – Volume 5)

CARRÉ Dominique, VIDAL Geneviève
Hyperconnectivity: Economical, Social and Environmental Challenges
(Computing and Connected Society Set – Volume 3)

CHAMOUX Jean-Pierre
The Digital Era 1: Big Data Stakes

DOUAY Nicolas
Urban Planning in the Digital Age
(Intellectual Technologies Set – Volume 6)

FABRE Renaud, BENSOUSSAN Alain
The Digital Factory for Knowledge: Production and Validation of Scientific Results

GAUDIN Thierry, LACROIX Dominique, MAUREL Marie-Christine, POMEROL Jean-Charles
Life Sciences, Information Sciences

GAYARD Laurent
Darknet: Geopolitics and Uses
(Computing and Connected Society Set – Volume 2)

IAFRATE Fernando
Artificial Intelligence and Big Data: The Birth of a New Intelligence
(Advances in Information Systems Set – Volume 8)

LE DEUFF Olivier
Digital Humanities: History and Development
(Intellectual Technologies Set – Volume 4)

MANDRAN Nadine
Traceable Human Experiment Design Research: Theoretical Model and Practical Guide
(Advances in Information Systems Set – Volume 9)

PIVERT Olivier
NoSQL Data Models: Trends and Challenges

ROCHET Claude
Smart Cities: Reality or Fiction

SALEH Imad, AMMI, Mehdi, SZONIECKY Samuel
Challenges of the Internet of Things: Technology, Use, Ethics
(Digital Tools and Uses Set – Volume 7)

SAUVAGNARGUES Sophie
Decision-making in Crisis Situations: Research and Innovation for Optimal Training

SEDKAOUI Soraya
Data Analytics and Big Data

SZONIECKY Samuel
Ecosystems Knowledge: Modeling and Analysis Method for Information and Communication
(Digital Tools and Uses Set – Volume 6)

2017

BOUHAÏ Nasreddine, SALEH Imad
Internet of Things: Evolutions and Innovations
(Digital Tools and Uses Set – Volume 4)

DUONG Véronique
Baidu SEO: Challenges and Intricacies of Marketing in China

LESAS Anne-Marie, MIRANDA Serge
The Art and Science of NFC Programming
(Intellectual Technologies Set – Volume 3)

LIEM André
Prospective Ergonomics
(Human-Machine Interaction Set – Volume 4)

MARSAULT Xavier
Eco-generative Design for Early Stages of Architecture
(Architecture and Computer Science Set – Volume 1)

REYES-GARCIA Everardo
The Image-Interface: Graphical Supports for Visual Information
(Digital Tools and Uses Set – Volume 3)

REYES-GARCIA Everardo, BOUHAÏ Nasreddine
Designing Interactive Hypermedia Systems
(Digital Tools and Uses Set – Volume 2)

SAÏD Karim, BAHRI KORBI Fadia
Asymmetric Alliances and Information Systems:Issues and Prospects
(Advances in Information Systems Set – Volume 7)

SZONIECKY Samuel, BOUHAÏ Nasreddine
Collective Intelligence and Digital Archives: Towards Knowledge Ecosystems
(Digital Tools and Uses Set – Volume 1)

2016

BEN CHOUIKHA Mona
Organizational Design for Knowledge Management

BERTOLO David
Interactions on Digital Tablets in the Context of 3D Geometry Learning
(Human-Machine Interaction Set – Volume 2)

BOUVARD Patricia, SUZANNE Hervé
Collective Intelligence Development in Business

EL FALLAH SEGHROUCHNI Amal, ISHIKAWA Fuyuki, HÉRAULT Laurent, TOKUDA Hideyuki
Enablers for Smart Cities

FABRE Renaud, in collaboration with MESSERSCHMIDT-MARIET Quentin, HOLVOET Margot
New Challenges for Knowledge

GAUDIELLO Ilaria, ZIBETTI Elisabetta
Learning Robotics, with Robotics, by Robotics
(Human-Machine Interaction Set – Volume 3)

HENROTIN Joseph
The Art of War in the Network Age
(Intellectual Technologies Set – Volume 1)

KITAJIMA Munéo
Memory and Action Selection in Human–Machine Interaction
(Human–Machine Interaction Set – Volume 1)

LAGRAÑA Fernando
E-mail and Behavioral Changes: Uses and Misuses of Electronic Communications

LEIGNEL Jean-Louis, UNGARO Thierry, STAAR Adrien
Digital Transformation
(Advances in Information Systems Set – Volume 6)

NOYER Jean-Max
Transformation of Collective Intelligences
(Intellectual Technologies Set – Volume 2)

VENTRE Daniel
Information Warfare – 2nd edition

VITALIS André
The Uncertain Digital Revolution
(Computing and Connected Society Set – Volume 1)

2015

ARDUIN Pierre-Emmanuel, GRUNDSTEIN Michel, ROSENTHAL-SABROUX Camille
Information and Knowledge System
(Advances in Information Systems Set – Volume 2)

BÉRANGER Jérôme
Medical Information Systems Ethics

BRONNER Gérald
Belief and Misbelief Asymmetry on the Internet

IAFRATE Fernando
From Big Data to Smart Data
(Advances in Information Systems Set – Volume 1)

KRICHEN Saoussen, BEN JOUIDA Sihem
Supply Chain Management and its Applications in Computer Science

NEGRE Elsa
Information and Recommender Systems
(Advances in Information Systems Set – Volume 4)

POMEROL Jean-Charles, EPELBOIN Yves, THOURY Claire
MOOCs

SALLES Maryse
Decision-Making and the Information System
(Advances in Information Systems Set – Volume 3)

SAMARA Tarek
ERP and Information Systems: Integration or Disintegration
(Advances in Information Systems Set – Volume 5)

2014

DINET Jérôme
Information Retrieval in Digital Environments

HÉNO Raphaële, CHANDELIER Laure
3D Modeling of Buildings: Outstanding Sites

KEMBELLEC Gérald, CHARTRON Ghislaine, SALEH Imad
Recommender Systems

MATHIAN Hélène, SANDERS Lena
Spatio-temporal Approaches: Geographic Objects and Change Process

PLANTIN Jean-Christophe
Participatory Mapping

VENTRE Daniel
Chinese Cybersecurity and Defense

2013

BERNIK Igor
Cybercrime and Cyberwarfare

CAPET Philippe, DELAVALLADE Thomas
Information Evaluation

LEBRATY Jean-Fabrice, LOBRE-LEBRATY Katia
Crowdsourcing: One Step Beyond

SALLABERRY Christian
Geographical Information Retrieval in Textual Corpora

2012

BUCHER Bénédicte, LE BER Florence
Innovative Software Development in GIS

GAUSSIER Eric, YVON François
Textual Information Access

STOCKINGER Peter
Audiovisual Archives: Digital Text and Discourse Analysis

VENTRE Daniel
Cyber Conflict

2011

BANOS Arnaud, THÉVENIN Thomas
Geographical Information and Urban Transport Systems

DAUPHINÉ André
Fractal Geography

LEMBERGER Pirmin, MOREL Mederic
Managing Complexity of Information Systems

STOCKINGER Peter
Introduction to Audiovisual Archives

STOCKINGER Peter
Digital Audiovisual Archives

VENTRE Daniel
Cyberwar and Information Warfare

2010

BONNET Pierre
Enterprise Data Governance

BRUNET Roger
Sustainable Geography

CARREGA Pierre
Geographical Information and Climatology

CAUVIN Colette, ESCOBAR Francisco, SERRADJ Aziz
Thematic Cartography – 3-volume series
Thematic Cartography and Transformations – Volume 1
Cartography and the Impact of the Quantitative Revolution – Volume 2
New Approaches in Thematic Cartography – Volume 3

LANGLOIS Patrice
Simulation of Complex Systems in GIS

MATHIS Philippe
Graphs and Networks – 2nd edition

THERIAULT Marius, DES ROSIERS François
Modeling Urban Dynamics

2009

BONNET Pierre, DETAVERNIER Jean-Michel, VAUQUIER Dominique
Sustainable IT Architecture: the Progressive Way of Overhauling Information Systems with SOA

PAPY Fabrice
Information Science

RIVARD François, ABOU HARB Georges, MERET Philippe
The Transverse Information System

ROCHE Stéphane, CARON Claude
Organizational Facets of GIS

2008

BRUGNOT Gérard
Spatial Management of Risks

FINKE Gerd
Operations Research and Networks

GUERMOND Yves
Modeling Process in Geography

KANEVSKI Michael
Advanced Mapping of Environmental Data

MANOUVRIER Bernard, LAURENT Ménard
Application Integration: EAI, B2B, BPM and SOA

PAPY Fabrice
Digital Libraries

2007

DOBESCH Hartwig, DUMOLARD Pierre, DYRAS Izabela
Spatial Interpolation for Climate Data

SANDERS Lena
Models in Spatial Analysis

2006

CLIQUET Gérard
Geomarketing

CORNIOU Jean-Pierre
Looking Back and Going Forward in IT

DEVILLERS Rodolphe, JEANSOULIN Robert
Fundamentals of Spatial Data Quality

Printed and bound by CPI Group (UK) Ltd, Croydon, CR0 4YY

15/09/2026

14962347-0004